by Adrian Harrison

INTRODUCTION TO DERRIVATIVES

January 2020

Contents

THE DERIVATIVE
Definition:

$$\lim_{x \to x_0} \frac{f(x) - f(x_0)}{x - x_0}$$

This expression is called derivative of $f(x)$ *at point* x_0

$$f'(x) = \frac{df(x)}{dx} = \frac{dy}{dx}$$

The derivative of $f(x)$ *at* $x = x_0$ *can be shown as*

$$f'(x_0), \frac{df(x_0)}{dx}, \frac{dy(x_0)}{dx}, y'(x_0)$$

☞ $x - x_0 = h$

$\Rightarrow x \to x_0 \Leftrightarrow h \to 0$

$$f'(x_0) = \lim_{h \to 0} \frac{f(x_0 + h) - f(x_0)}{h}$$

☞ $f'(x) = \dfrac{df(x)}{dx} = \dfrac{dy}{dx} = \lim_{h \to 0} \dfrac{f(x + h) - f(x)}{h}$

Example:

$f : R \to R,$

$f'(x) = x^3 - 4x^2 + 2x - 3 \Rightarrow f'(2) = ?$

Solution:

$$f'(2) = \lim_{x \to 2} \frac{f(x) - f(2)}{x - 2} = \lim_{x \to 2} \frac{x^3 - 4x^2 + 2x - 3 + 7}{x - 2}$$

1

$$f'(2) = \lim_{x \to 2} \frac{x^3 - 4x^2 + 2x + 4}{x - 2}$$

$$= \lim_{x \to 2} \frac{(x - 2)(x^2 - 2x - 2)}{x - 2}$$

$$f'(2) = \lim_{x \to 2} (x^2 - 2x - 2) = 4 - 4 - 2 = -2$$

$$f'(2) = -2$$

Example:

$$f{:}R \to R, \ f(x) = x^2 - 4x + 4 \Rightarrow f'(x) = \frac{df(x)}{dx} = \ ?$$

Solution:

$$f'(x) = \lim_{h \to 0} \frac{f(x + h) - f(x)}{h}$$

$$f'(x) = \lim_{h \to 0} \frac{(x^2 + 2hx + h^2) - (4x + 4h) + 4 - x^2 + 4x - 4}{h}$$

$$f'(x) = \lim_{h \to 0} \frac{h(2x - 4 + h)}{h} = 2x - 4$$

$$f'(x) = x^2 - 4x + 4 \Rightarrow f'(x) = 2x - 4$$

RULES FOR TAKING DERIVATIVE

1. $f(x) = c \Rightarrow f'(x) = 0 \ (c \in R)$

2. $f(x) = x^n \Rightarrow f'(x) = n \cdot x^{n-1}$

3. $(f(x) \cdot g(x))' = (f'(x) \cdot g(x)) + (f(x) \cdot g'(x))$

4. $(f(x) \pm g(x))' = f'(x) \pm g'(x)$

5. $\left(\dfrac{f(x)}{g(x)}\right)' = \dfrac{f'(x) \cdot g(x) - f(x) \cdot g'(x)}{(g(x))^2}$

6. $(k \cdot f(x))' = k \cdot f'(x) \ (k \in R)$

7. $(f^m(x))' = m \cdot f^{m-1}(x) \cdot f'(x)$

8. $\left(\sqrt[n]{f(x)}\right)' = \dfrac{f'(x)}{n \cdot \sqrt[n]{f^{n-1}(x)}}$

9. $\left(\sqrt{f(x)}\right)' = \dfrac{f'(x)}{2 \cdot \sqrt{f(x)}}$

10. $|f(x)|' = \dfrac{f'(x) \cdot |f(x)|}{f(x)}$

$|f(x)|' = \begin{cases} f'(x), & f(x) > 0 \\ -f'(x), & f(x) < 0 \end{cases}$

11. $(f(u(x))' = u'(x) \cdot f'(u(x))$

Examples:

1. $f(x) = 5 \Rightarrow f'(x) = 0$

2. $f(x) = x \Rightarrow f'(x) = 1$

3. $f(x) = 7x \Rightarrow f'(x) = 7$

4. $f(x) = x^7 \Rightarrow f'(x) = 7x^6$

5. $f(x) = -3x^5 \Rightarrow f'(x) = -15x^4$

6. $f(x) = 2x^3 - 5x^2 + 6x - 7 \Rightarrow f'(x) = 6x^2 - 10x + 6$

7. $f(x) = (x^2 + 2) \cdot (x^3 + x + 1)$

$f'(x) = 2x \cdot (x^3 + x + 1) + (x^2 + 2) \cdot (3x^2 + 1)$

8. $f(x) = \dfrac{x}{x^2 + 3} \Rightarrow f'(x) = \dfrac{1(x^2 + 3) - 2x(x)}{(x^2 + 3)^2}$

$= \dfrac{x^2 + 3 - 2x^2}{(x^2 + 3)^2} = \dfrac{3 - x^2}{(x^2 + 3)^2}$

9. $f(x) = \sqrt[5]{x^2 + 2x + 3} \Rightarrow f(x) = (x^2 + 2x + 3)^{1/5}$

$\Rightarrow f'(x) = (2x + 2) \cdot \dfrac{1}{5}(x^2 + 2x + 3)^{-4/5}$

$\Rightarrow f'(x) = \dfrac{2(x + 1)}{5 \cdot \sqrt[5]{(x^2 + 2x + 3)^4}}$

10. $f(x) = |x^2 - 5x + 6|$

$f'(2)$ *and* $f'(3)$ *do not exist because* $g(2) = 0$ *and* $g(3) = 0$

$f'(1) = ? \Rightarrow f'(x) = 2x - 5 = -3$

DERIVATIVE OF CLOSED FUNCTIONS

$$F(x,y) = 0 \; \Rightarrow \; \frac{dy}{dx} = -\frac{F'_x(x,y)}{F'_y(x,y)}$$

$F'x$:*derivative of F with respect to x*

$F'y$:*derivative of F with respect to y*

Example:

$$F(x,y) = x^4 \cdot y^3 + 2x^3 + 4xy = 0 \Rightarrow y = \frac{dy}{dx} = \; ?$$

Solution:

$$4x^3 \cdot y^3 + 3y^2 \cdot x^4 \cdot y' + 6x^2 + 4y + 4x \cdot y' = 0$$

$$y'(3x^4 \cdot y^2 + 4x) = -(4x^3 \cdot y^3 + 6x^2 + 4y)$$

$$y' = -\frac{4x^3 \cdot y^3 + 6x^2 + 4y}{3x^4 \cdot y^2 + 4x}$$

DERIVATIVE OF COMBINING FUNCTIONS

$$(gof)'(x) = g'(f(x)) \cdot f'(x)$$

Example:

$$f(x) = 4x^2 + 2$$

$$g(x) = x^3 - 3$$

$$\Rightarrow (gof)'(x) = ?$$

Solution:

$$(gof)'(x) = \left(4x^2 + 2\right)^3 - 3$$

$$(gof)'(x) = 3\left(4x^2 + 2\right)^2 \cdot 8x$$

$$= 12(2x + 1)^2 \cdot 8x$$

$$= 96x \cdot (2x + 1)^2$$

DERIVATIVE OF PARAMETRIC FUNCTIONS

$$\begin{cases} x = f(t) \\ y = g(t) \end{cases} \Rightarrow \frac{dy}{dx} = \frac{dy}{dt} \cdot \frac{dt}{dx} = \frac{\dfrac{dy}{dt}}{\dfrac{dx}{dt}}$$

Example:

$$\begin{cases} x = 6t - 3t^2 \\ y = 4t^3 + 3t^2 \end{cases} \Rightarrow \frac{dy}{dx}\bigg|_{t=2} = \ ?$$

Solution:

$$f'(x) = \frac{dy}{dx} = \frac{dy}{dt} \cdot \frac{dt}{dx}$$

$$f'(x) = (12t^2 + 6t) \cdot \frac{1}{6 - 6t}$$

$$f'(x) = 6(2t^2 + t) \cdot \frac{1}{6(1 - t)}$$

$$t = 2 \Rightarrow$$

$$x = 12 - 12 = 0$$

$$f'(0) = (8 + 2) \cdot \frac{1}{6(-1)} = -\frac{5}{3}$$

☞ $$\begin{cases} x = f(t) \\ y = g(t) \end{cases} \Rightarrow \frac{d^2y}{dx^2} = \frac{d^2y}{dx^2} = \frac{d}{dt}\left(\frac{dy}{dx}\right) \cdot \frac{dt}{dx}$$

Example:

$$\begin{cases} x = 3t^2 + 3t \\ y = t^3 - 3t \end{cases} \Rightarrow \frac{d^2y}{dx^2}\bigg|_{t=1} = ?$$

Solution:

$$\frac{dy}{dx} = \frac{\dfrac{dy}{dt}}{\dfrac{dx}{dt}} = \frac{3t^2 - 3}{6t + 3} = \frac{t^2 - 1}{2t + 1}$$

$$\frac{d^2y}{dx^2} = \frac{d}{dt}\left(\frac{dy}{dx}\right) \cdot \frac{dt}{dx}$$

$$\frac{d^2y}{dx^2} = \frac{2t(2t + 1) - 2(t^2 - 1)}{(2t + 1)^2} \cdot \frac{1}{6t + 3}$$

$$t = 1 \Rightarrow \frac{d^2y}{dx^2} = \frac{2 \cdot 3 - 2 \cdot 0}{(3)^2} \cdot \frac{1}{9}$$

$$\frac{d^2y}{dx^2} = \frac{6}{81} = \frac{2}{27}$$

DERIVATIVE OF TRIGONOMETRIC FUNCTIONS

$f(x) = \sin x \Rightarrow f'(x) = \cos x$

$f(x) = \sin(u(x)) \Rightarrow f'(x) = u'(x) \cdot \cos(u(x))$

$f(x) = \sin^n(u(x)) \Rightarrow f'(x) = n \cdot u'(x) \cdot \sin^{n-1}(u(x)) \cdot \cos(u(x))$

$f(x) = \cos x \Rightarrow f'(x) = -\sin x$

$f(x) = \cos(u(x)) \Rightarrow f'(x) = -u'(x) \cdot \sin(u(x))$

$f(x) = \cos^n(u(x)) \Rightarrow f'(x) = -n(u'(x) \cdot \cos^{n-1}(u(x)) \cdot \sin(u(x))$

Example:

1. $f(x) = \sin^3(x^2 + x)$

$$\Rightarrow f'(x) = 3(2x + 1) \cdot \sin^2(x^2 + x) \cdot \cos(x^2 + x)$$

2. $f(x) = \cos^4(3x) \Rightarrow f'(x) = -4 \cdot 3 \cos^3(3x) \cdot \sin(3x)$

$$\Rightarrow f'(x) = -12 \cos^3(3x) \cdot \sin(3x)$$

$f(x) = \tan x \Rightarrow f'(x) = (1 + \tan^2 x) = \sec^2 x$

$$f(x) = \tan(u(x)) \Rightarrow f'(x) = u'(x)(1 + tan^2(u(x)))$$

$$= u'(x) \cdot sec^2(u(x))$$

$$f(x) = tan^n(u(x))$$

$$f'(x) = n \cdot u'(x) \cdot tan^{n-1}(u(x))\,(1 + tan^2(u(x)))$$

$$= n \cdot u'(x)\, tan^{n-1}(u(x)) \cdot sec^2(u(x))$$

Examples:

1. $f(x) = \tan 6x \Rightarrow f'(x) = 6(1 + tan^2 6x) = 6\,sec^2 6x$

2. $f(x) = tan^3(x^2 - 1)$

$\Rightarrow f'(x) = 3 \cdot 2x\, tan^2(x^2 - 1) \cdot (1 + tan^2(x^2 - 1))$

$\Rightarrow f'(x) = 6x\, tan^2(x^2 - 1) \cdot sec^2(x^2 - 1)$

$f(x) = \cot x \Rightarrow f'(x) = -(1 + cot^2 x) = -cosec^2 x$

$f(x) = \cot(u(x)) \Rightarrow f'(x) = -u'(x)(1 + cot^2(u(x)))$

$f'(x) = -u'(x) \cdot cossec^2(u(x))$

$f(x) = cot^n(u(x)) \Rightarrow f'(x) =$

$-n(u'(x))\, cot^{n-1}(u(x)) \cdot (1 + cot^2(u(x)))$

$\Rightarrow f'(x) = -n \cdot (u'(x))\, cot^{n-1}(u(x)) \cdot cosec^2(u(x))$

Examples:

1. $f(x) = \cot 4x \Rightarrow f'(x) = -4(1 + \cot^2 4x) = -4\,cosec^2\,4x$

2. $f(x) = \cot^2(x^2 + 3x)$

$\Rightarrow f'(x) = -2(2x + 3) \cdot \cot(x^2 + 3x)\left(1 + \cot^2(x^2 + 3x)\right)$

$\Rightarrow f'(x) = -2(2x + 3) \cdot \cot(x^2 + 3x) \cdot cosec^2(x^2 + 3x)$

DERIVATIVE OF INVERSE TRIGONOMETRIC FUNCTIONS

$f(x) = \arcsin{x}$ $\qquad \Rightarrow f'(x) = \dfrac{1}{\sqrt{1-x^2}}$

$f(x) = \arcsin(u(x))$ $\qquad \Rightarrow f'(x) = \dfrac{u'(x)}{\sqrt{1-(u(x))^2}}$

$f(x) = \arccos{x}$ $\qquad \Rightarrow f'(x) = \dfrac{-1}{\sqrt{1-x^2}}$

$f(x) = \arccos(u(x))$ $\qquad \Rightarrow f'(x) = \dfrac{-u'(x)}{\sqrt{1-(u(x))^2}}$

$f(x) = \arctan{x}$ $\qquad \Rightarrow f'(x) = \dfrac{1}{1+x^2}$

$f(x) = \arctan(u(x))$ $\qquad \Rightarrow f'(x) = \dfrac{u'(x)}{1+(u(x))^2}$

$f(x) = \text{arccot}\,x$ $\qquad \Rightarrow f'(x) = \dfrac{-1}{1+x^2}$

$f(x) = \text{arccot}(u(x))$ $\qquad \Rightarrow f'(x) = \dfrac{-u'(x)}{1+(u(x))^2}$

Examples:

1. $f(x) = \arctan(x^2 + 2x) - \arccos(2x) \Rightarrow f'(0) = ?$

12

$$f'(x) = \frac{2x + 2}{1 + (x^2 + 2x)^2} - \frac{-2x}{\sqrt{1 - (2x)^2}}$$

$$f'(x) = \frac{2(x + 1)}{1 + (x^2 + 2x)^2} + \frac{2x}{\sqrt{1 - 4x^2}}$$

$$f'(0) = \frac{2(0 + 1)}{1 + 0^2} + \frac{2 \cdot 0}{\sqrt{1 - 0}} \Rightarrow f'(0) = 2$$

2. $f(x) = arccot(x^2 + 2x) - (arcsin\, x)^2 \Rightarrow f'\left(\frac{1}{2}\right) = ?$

$$f'(x) = \frac{-(2x + 2)}{1 + (x^2 + 2x)^2} - \frac{1}{\sqrt{1 - x^2}} \cdot 2\, arcsin\, x$$

$$f'\left(\frac{1}{2}\right) = \frac{-\left(2 \cdot \frac{1}{2} + 2\right)}{1 + \left(\frac{1}{4} + 2 \cdot \frac{1}{2}\right)^2} - \frac{1}{\sqrt{1 - \frac{1}{4}}} \cdot 2\, arcsin\frac{1}{2}$$

$$f'\left(\frac{1}{2}\right) = \frac{-3}{1 + \frac{25}{16}} - \frac{2}{\sqrt{3}} \cdot 2 \cdot \frac{\pi}{6}$$

$$= \frac{-48}{41} - \frac{2\sqrt{3}\,\pi}{9} = -\frac{432 + 82\sqrt{3}\,\pi}{369}$$

DERIVATIVE OF LOGARITHMIC FUNCTIONS

$$f(x) = \log_a x \Rightarrow f'(x) = \frac{1}{x \cdot \ln a} = \frac{1}{x} \cdot \log_a e$$

$$f(x) = \log_a (u(x)) \Rightarrow f'(x) = \frac{u'(x)}{u(x) \cdot \ln a} = \frac{u'(x)}{u(x)} \cdot \log_a e$$

$$f(x) = \ln x \Rightarrow f'(x) = \frac{1}{x}$$

$$f(x) = \ln (u(x)) \Rightarrow f'(x) = \frac{u'(x)}{u(x)}$$

Examples:

1. $f(x) = \log_5 (x^2 + 4x) \Rightarrow f'(2) = ?$

$$f'(x) = \frac{2x + 4}{x^2 + 4x} \log_5 e \Rightarrow f'(2) = \frac{8}{12} \cdot \log_5 e$$

$$\Rightarrow f'(2) = \frac{2}{3} \cdot \log_5 e$$

2. $f(x) = \ln (x^3 + 6x) \Rightarrow f'(1) = ?$

$$f'(x) = \frac{3x^2 + 6}{x^3 + 6x} \Rightarrow f'(1) = \frac{3 \cdot 1 + 6}{1 + 6} = \frac{9}{7}$$

DERIVATIVE OF EXPONENTIAL FUNCTIONS

$$f(x) = e^x \Rightarrow f'(x) = e^x$$

$$f(x) = e^{u(x)} \Rightarrow f'(x) = u'(x) \cdot e^{u(x)}$$

$$f(x) = a^x \Rightarrow f'(x) = a^x \cdot \ln a = \frac{1}{\log_a e} \cdot a^x$$

$$f(x) = a^{u(x)} \Rightarrow f'(x) = u'(x) \cdot a^{u(x)} \cdot \ln a = \frac{u'(x)}{\log_a e} a^{u(x)}$$

Examples:

$1. f(x) = e^{\ln x} \Rightarrow f'(e) = ?$

$$f(x) = e^{\ln x} \Rightarrow f'(x) = \frac{1}{x} \cdot e^{\ln x}$$

$$f'(e) = \frac{1}{e} \cdot e^{\ln e} = \frac{1}{e} \cdot e = 1$$

$2. f(x) = 5^{x^2 + 4} \Rightarrow f'(x) = ?$

$$f'(x) = 2x \cdot 5^{x^2 + 4} \cdot \ln 5$$

HIGHER ORDER DERIVATIVES

$y' = f'(x)$ 1. 1st order derivative

$y'' = f''(x)$ 2. 2nd order derivative

$y''' = f'''(x)$ 3. 3rd order derivative

.

.

.

$y^{(n)} = f^{(n)}(x)$ n. n^{th} order derivative

Example:

$f(x) = x^3 + 4x^2 - 2x + 6 \Rightarrow f'''(x) = ?$

Solution:

$f'(x) = 3x^2 + 8x - 2 \Rightarrow f''(x) = 6x + 8 \Rightarrow f'''(x) = 6$

Example:

$$f(x,y) = x^2 + y^2 - 9 = 0 \Rightarrow f''(x,y) = ?$$

Solution:

$$y' = f'(x) = \frac{-2x}{2y} = \frac{-x}{y}$$

$$y'' = f''(x) = \frac{-1 \cdot y - xy'}{y^2}$$

$$= \frac{-y + x\left(\dfrac{-x}{y}\right)}{y^2} = \frac{-y^2 - x^2}{y^3}$$

$$= -\frac{x^2 + y^2}{y^3}$$

L´ HOSPITAL RULE

If $\lim\limits_{x \to x_0} \dfrac{f(x)}{g(x)}$ *is equal to* $\dfrac{0}{0}$ *or* $\dfrac{\infty}{\infty}$ *, derivatives of*

numerator and denominator

are taken separately

$$\lim\limits_{x \to x_0} \dfrac{f(x)}{g(x)} = \lim\limits_{x \to x_0} \dfrac{f'(x_0)}{g'(x_0)}$$

If the limit $\lim\limits_{x \to x_0} \dfrac{f'(x_0)}{g'(x_0)}$ *regivesthe same uncertainity*

, apply the rule again,

that is, take the 2nd *derivative.*

Examples:

1. $\lim\limits_{x\to 8}\dfrac{\sqrt{x}-2\sqrt{2}}{\sqrt[3]{x}-2}=\lim\limits_{x\to 8}\dfrac{\sqrt{8}-2\sqrt{2}}{\sqrt[3]{x}-2}$

$$=\dfrac{2\sqrt{2}-2\sqrt{2}}{2-2}=\dfrac{0}{0}$$

$\lim\limits_{x\to 8}\dfrac{(\sqrt{x}-2\sqrt{2})}{(\sqrt[3]{x}-2)}=\lim\limits_{x\to 8}\dfrac{\dfrac{1}{2\sqrt{x}}}{\dfrac{1}{3\cdot\sqrt[3]{x^2}}}$

$$=\dfrac{1}{2\sqrt{8}}\cdot\dfrac{3\sqrt[3]{8}}{1}=\dfrac{3\sqrt{2}}{4}$$

2. $\lim\limits_{x\to 0}=\dfrac{e^{\ln x}-e}{\ln^{[ro]}(\ln x)}$

$$=\lim\limits_{x\to 0}\dfrac{e^{\ln x}-e}{\ln^{[ro]}(\ln x)}$$

$$=\dfrac{e-e}{\ln^{[ro]}(\ln e)}=\dfrac{0}{\ln 1}=\dfrac{0}{0}$$

$\lim\limits_{x\to 0}\dfrac{e^{\ln x}-e}{\ln^{[ro]}(\ln x)}$

$$=\lim\limits_{x\to 0}\dfrac{\dfrac{1}{x}e^{\ln x}}{\dfrac{1}{x}\cdot\dfrac{1}{\ln x}}=\dfrac{\dfrac{1}{e}}{\dfrac{1}{e}\cdot 1}=e$$

TEST WITH SOLUTIONS

1. $f(x) = 2x^3 - 5x^2 + 4x - 7 \implies f'(x) = ?$

A) $3x^2 - 5x - 4$ B) $6x^2 - 10x + 4$ C) $6x^2 - 10x - 7$

D) $3x^2 - 5x + 4$ E) $x^3 - x^2 + 4$

Solution:

$f(x) = 2x^3 - 5x^2 + 4x - 7$

$f(x) = 6x^2 - 10x + 4$

Correct Answer - B

2. $f(x) = (x^2 - 1) \cdot (2x^2 - 3x + 1) \implies f'(2) = ?$

A) 21   B) 24   C) 27   D) 30   E) 33

Solution:

$$f'(x) = (x^2 - 1)' \cdot (2x^2 - 3x + 1) + (2x^2 - 3x + 1)' \cdot (x^2 - 1)$$

$$= 2x \cdot (2x^2 - 3x + 1) + (4x - 3) \cdot (x^2 - 1)$$

$$f'(2) = 2 \cdot 2(2 \cdot 2^2 - 3 \cdot 2 + 1) + (4 \cdot 2 - 3) \cdot (2^2 - 1)$$

$$= 4 \cdot (8 - 6 + 1) + (8 - 3) \cdot (4 - 1)$$

$$= 4 \cdot 3 + 5 \cdot 3$$

$$= 12 + 15$$

$$= 27$$

Correct Answer - C

3. $f(x) = \dfrac{x^2 - 3}{3x + 1} \Rightarrow f'(-1) = ?$

A) 1   B)$\dfrac{3}{2}$ C) 2   D)$\dfrac{5}{2}$ E) 3

Solution:

$$f'(x) = \frac{(x^2 - 3)' \, (3x + 1) - (3x + 1)' \cdot (x^2 - 3)}{(3x + 1)}$$

$$f'(x) = \frac{2 \cdot x(3x + 1) - 3 \cdot (x^2 - 3)}{(3x + 1)^2}$$

$$f'(x) = \frac{6x^2 + 2x - 3x^2 + 9}{(3x + 1)^2}$$

$$f'(-1) = \frac{3 \cdot (-1)^2 + 2 \cdot (-1) + 9}{(3 \cdot (-1) + 1)^2}$$

21

$$= \frac{3 - 2 + 9}{(-2)^2}$$

$$= \frac{10}{4}$$

$$= \frac{5}{2}$$

Correct Answer - D

4. $f(x) = \dfrac{x^3 - 1}{x} \Rightarrow f'(x) = \ ?$

A) $3x + \dfrac{1}{x}$ B) $2x + \dfrac{1}{x^2}$ C) $3x - \dfrac{1}{x^2}$

D) $4x + \dfrac{1}{x^2}$ E) $3x + \dfrac{1}{x^2}$

Solution:

$$f(x) = \frac{x^3 - 1}{x}$$

$$f(x) = \frac{x^3}{x} - \frac{1}{x}$$

$$f(x) = x^2 - x^{-1}$$

$$f'(x) = 2x + x^{-2}$$

$$f'(x) = 2x + \frac{1}{x^2}$$

Correct Answer - B

5. $f(x) = (x^4 - 2x)^5 \Rightarrow f'(1) = ?$

   A) 10   B) 12   C) 14   D) 16   E) 18

Solution:

$f(x) = (x^4 - 2x)^5$

$f'(x) = 5 \cdot (x^4 - 2x)^4 \cdot (x^4 - 2x)'$

$\qquad = 5 \cdot (x^4 - 2x)^4 \cdot (4x^3 - 2)$

$f'(1) = 5 \cdot (1^4 - 2 \cdot 1)^4 \cdot (4 \cdot 1^3 - 2)$

$\qquad = 5 \cdot (-1)^4 \cdot 2$

$\qquad = 5 \cdot 1 \cdot 2$

$\qquad = 10$

Correct Answer - A

6. $f(x) = \sqrt[3]{x^2 - 3} \Rightarrow f'(2) = ?$

   A) $\dfrac{1}{5}$ B) $\dfrac{4}{3}$ C) $\dfrac{1}{2}$ D) 1   E) 2

Solution:

$f(x) = \sqrt[3]{x^2 - 3} = (x^2 - 3)^{1/3}$

$f'(x) = \dfrac{1}{3}(x^2 - 3)^{-2/3} \cdot (x^2 - 3)'$

23

$$f'(x) = \frac{1}{3}(x^2 - 3)^{-2/3} \cdot (2x)$$

$$f'(2) = \frac{1}{3} \cdot (2^2 - 3)^{-2/3} \cdot 2 \cdot 2$$

$$= \frac{1}{3} \cdot 1 \cdot 2 \cdot 2$$

$$= \frac{4}{3}$$

Correct Answer - B

7. $x^2 + 2xy - y^2 = 0 \quad \Rightarrow \frac{dy}{dx} = ?$

A) $\dfrac{x - y}{y}$ B) $\dfrac{y - x}{x + y}$ C) $\dfrac{x + y}{y - x}$

D) $\dfrac{x + y}{x - y}$ E) $\dfrac{x - y}{x + y}$

Solution:

$$x^2 + 2xy - y^2 = 0$$

$$\frac{dy}{dx} = -\frac{2x + 2y}{2x - 2y}$$

$$= -\frac{2 \cdot (x + y)}{2 \cdot (x - y)}$$

$$= \frac{x + y}{y - x}$$

8. $f(4) = -3, f'(4) = 2$ *and* $g'(-3) = -5$

$\Rightarrow (gof)'(4) = ?$

A) -16 B) -14 C) -12 D) -10 E) -8

Solution:

$(gof)'(x) = g'(f(x)) \cdot f'(x)$

$(gof)'(4) = g'(f(4)) \cdot f'(4)$

$= g'(-3) \cdot 2$

$= (-5) \cdot 2$

$= -10$

Correct Answer - D

9. $y = 3x^2 - 1, z = 2y^3 + 4 \Rightarrow \dfrac{dz}{dx} = ?$

A) $36x(3x^2 - 1)^2$ B) $6x \cdot (3x^2 - 1)^2$ C) $36 \cdot (x^2 - 1)^2$

D) $x(3x^2 - 1)^2$ E) $18x(3x^2 - 1)$

Solution:

$\dfrac{dz}{dx} = \dfrac{dz}{dy} \cdot \dfrac{dy}{dx}$

$= 6y^2 \cdot 6x$

$$= 36 \cdot y^2 \cdot x$$

$$= 36x \cdot (3x^2 - 1)^2 \cdot x$$

Correct Answer - A

10. $\begin{cases} x = 4t - t^2 \\ y = 2t^2 + t \end{cases} \Rightarrow \dfrac{dy}{dx} = ?$

A) $\dfrac{2t + 4}{4t - 2}$ B) $\dfrac{4t - 1}{2t + 4}$ C) $\dfrac{4t + 1}{4 - 2t}$

D) $\dfrac{2t - 4}{4t + 1}$ E) $\dfrac{t^2 - 1}{4 + 2t}$

Solution:

$$\frac{dy}{dx} = \frac{\dfrac{dy}{dt}}{\dfrac{dx}{dt}}$$

$$= \frac{4t + 1}{4 - 2t}$$

Correct Answer - C

11. $f(x) = sin(3x + 2) \Rightarrow f'(x) = ?$

A) $-2\,cos(3x + 2)$ B) $2\,cos(3x + 2)$ C) $2\,sin(3x + 2)$

D) $3 \cdot cos(3x + 2)$ E) $-3 \cdot cos(3x + 2)$

Solution:

$$f(x) = sin(3x + 2)$$

$$f'(x) = cos(3x + 2) \cdot 3$$

$$= 3 \cdot cos(3x + 2)$$

Correct Answer - D

12. $f(x) = cos^2(3x) \Rightarrow f'\left(\dfrac{\pi}{6}\right) = ?$

A) $-\dfrac{1}{4}$ B) $-\dfrac{1}{2}$ C) 0   D) $\dfrac{1}{2}$ E) $\dfrac{1}{4}$

Solution:

$$f(x) = cos^2(3x)$$

$$f'(x) = 2\,cos(3x) \cdot ((cos\ 3x))'$$

$$= 2 \cdot cos\ (3x) \cdot (-sin(3x) \cdot 3)$$

$$= -6 \cdot cos\ 3x \cdot sin\ 3x$$

$$= -3 \cdot 2 \cdot sin(3x) \cdot cos(3x),$$

$$sin\ 2\propto\ = 2 \cdot sin \propto cos \propto$$

$$= -3 \cdot sin(6x)$$

$$f'\left(\frac{\pi}{6}\right) = -3 \cdot sin\left(6 \cdot \frac{\pi}{6}\right)$$

$$= -3 \cdot sin(\pi)$$

$$= -3 \cdot 0$$

27

$$= 0$$

13. $f(x) = \tan 4x + \cot 2x \Rightarrow f'\left(\dfrac{2\pi}{3}\right) = \ ?$

A)$\dfrac{32}{3}$ B) 11   C)$\dfrac{35}{3}$ D) 13   E)$\dfrac{40}{3}$

Solution:

$f(x) = \tan 4x + \cot 2x$

$f'(x) = (1 + \tan^2 4x)\cdot 4 - (1 + \cot^2 2x)\cdot 2$

$f'\left(\dfrac{2\pi}{3}\right) = \left(1 + \tan^2\dfrac{8\pi}{3}\right)\cdot 4 - (1 + \cot^2 2x)\cdot 2$

$\quad = \left(1 + \tan^2\dfrac{2\pi}{3}\right)\cdot 4 - \left(1 + \cot^2\dfrac{4\pi}{3}\right)\cdot 2$

$\quad = \left(1 + (-\sqrt{3})^2\right)\cdot 4 - \left(1 + \left(\dfrac{1}{\sqrt{3}}\right)^2\right)\cdot 2$

$\quad = (1 + 3)\cdot 4 - \left(1 + \dfrac{1}{3}\right)\cdot 2$

$\quad = 16 - \dfrac{8}{3}$

$\quad = \dfrac{40}{3}$

Correct Answer - E

28

14. $f(x) = \sin^3 2x \quad \Rightarrow f'\left(\dfrac{\pi}{4}\right) = ?$

A) $-\dfrac{\sqrt{3}}{2}$ B) $-\dfrac{1}{2}$ C) 0   D) $\dfrac{1}{2}$ E) $\dfrac{\sqrt{3}}{2}$

Solution:

$f(x) = \sin^3 2x$

$f'(x) = 3 \cdot \sin^2 2x \cdot \cos 2x \cdot 2$

$f'\left(\dfrac{\pi}{4}\right) = 3 \cdot \sin^2 \dfrac{\pi}{2} \cdot \cos \dfrac{\pi}{2} \cdot 2$

$\qquad = 0$

Correct Answer - C

15. $f(x) = \arctan(x^2 - 1) \quad \Rightarrow f'(2) = ?$

A) $\dfrac{2}{5}$ B) $\dfrac{3}{5}$ C) $\dfrac{4}{5}$ D) 1   E) $\dfrac{6}{5}$

Solution:

$f'(x) = \dfrac{2x}{1 + (x^2 - 1)^2}$

$f'(2) = \dfrac{4}{1 + (2^2 - 1)^2}$

$\qquad = \dfrac{4}{10} = \dfrac{2}{5}$

16. $f(x) = \arctan x + (\arccos x)^2 \quad \Rightarrow \quad f'(0) = ?$

A) $1 + \pi$ B) $1 - \pi$ C) $2 + \pi$

D) $2 - \pi$ E) $3 + \pi$

Solution:

$$f(x) = \arctan x + (\arccos x)^2$$

$$f'(x) = \frac{1}{1 + x^2} + 2 \cdot (\arccos x) \cdot \left(-\frac{1}{\sqrt{1 + x^2}} \right)$$

$$f'(0) = \frac{1}{1 + 0} + 2 \cdot (\arccos 0) \cdot \left(-\frac{1}{\sqrt{1 + 0}} \right)$$

$$f'(0) = 1 + 2 \cdot \frac{\pi}{2} \cdot (-1)$$

$$= 1 - \pi$$

Correct Answer - B

17. $f(x) = \ln (x^2 + 3x) \quad \Rightarrow \quad f'(x) = ?$

A) $\dfrac{2x + 2}{x^2 + 2x}$ B) $\dfrac{2}{x + 3x^2}$ C) $\dfrac{2x + 3}{x^2 + 3x}$

D) $\dfrac{2x + 3}{x + 3}$ E) $\dfrac{2x}{x^2 + 3}$

Solution:

30

$$f'(x) = \frac{2x + 3}{x^2 + 3x}$$

Correct Answer - C

18. $f(x) = \ln{(\cos x)} \quad \Rightarrow \quad f'(x) = ?$

    A) $\sin x$ B) $\cot x$ C) $-\cot x$

    D) $-\tan x$ E) $\tan x$

Solution:

$f(x) = \ln{(\cos x)}$

$$f'(x) = \frac{-\sin x}{\cos x}$$

$f'(x) = -\tan x$

Correct Answer - D

19. $f(x) = \log_5(x^2 - 2) \quad \Rightarrow \quad f'(5) = ?$

    A)$\dfrac{13}{5 \ln 5}$ B)$\dfrac{23}{2 \cdot \ln 5}$ C)$\dfrac{13}{\ln 5}$

    D)$\dfrac{2}{5 \cdot \ln 5}$ E)$\dfrac{10}{23 \cdot \ln 5}$

Solution:

$f(x) = \log_5(x^2 - 2)$

$$f'(x) = \frac{2x}{(x^2 - 2) \cdot \ln 5}$$

$$f'(5) = \frac{2 \cdot 5}{(5^2 - 2) \cdot \ln 5}$$

$$= \frac{10}{23 \cdot \ln 5}$$

Correct Answer - E

20. $f(x) = \ln (cosec\, x + cot\, x) \;\Rightarrow\; f'(x) = ?$

A) $\dfrac{cot\, x + sin\, x}{sin\, x - 1}$ 　　B) $\dfrac{sin\, x}{sin\, x - 1}$ 　　C) $\dfrac{cos\, x}{cos\, x - 1}$

D) $sec\, x$ 　　　　　E) $- cosec\, x$

Solution:

$f(x) = \ln (cosec\, x + cot\, x)$

$$f(x) = \ln \left(\frac{1}{sin\, x} + \frac{cos\, x}{sin\, x} \right)$$

$$f(x) = \ln \left(\frac{1 + cos\, x}{sin\, x} \right)$$

$$f'(x) = \frac{\left(\dfrac{1 + cos\, x}{sin\, x} \right)'}{\dfrac{1 + cos\, x}{sin\, x}}$$

$$= \frac{\dfrac{-\sin x \cdot \sin x - \cos x \cdot (1 + \cos x)}{(\sin x)^2}}{\dfrac{1 + \cos x}{\sin x}}$$

$$= \frac{-\sin^2 x - \cos x - \cos^2 x}{\sin^2 x} \cdot \frac{\sin x}{1 + \cos x}$$

$$= \frac{(-1 - \cos x) \cdot \sin x}{\sin^2 x \cdot (1 + \cos x)}$$

$$= \frac{-(1 + \cos x) \cdot \sin x}{\sin^2 x (1 + \cos x)}$$

$$= -\frac{1}{\sin x}$$

$$= -\csc x$$

Correct Answer - E

21. $f(x) = e^{x^2 + 1} \quad \Rightarrow \quad f'(1) = ?$

A) $\dfrac{e^2}{2}$ B) $3e$ C) $2e^2$ D) $2e$ E) $3e^2$

Solution:

$f(x) = e^{x^2 + 1}$

$f'(x) = e^{x^2 + 1} \cdot 2x$

$f'(1) = e^2 \cdot 2$

$$f'(1) = 2 \cdot e^2$$

22. $f(x) = 5^{\cos x} \quad \Rightarrow f'\left(\frac{\pi}{2}\right) = ?$

A) $\ln\dfrac{1}{5}$ B) $\ln 5$ C) $\ln\dfrac{2}{5}$

D) $\ln 25$ E) $\ln\dfrac{3}{5}$

Solution:

$$f(x) = 5^{\cos x}$$

$$f'(x) = 5^{\cos x} \cdot (-\sin x) \cdot \ln 5$$

$$f'\left(\frac{\pi}{2}\right) = 5^{\cos\frac{\pi}{2}} \cdot \left(-\sin\frac{\pi}{2}\right) \cdot \ln 5$$

$$f'\left(\frac{\pi}{2}\right) = 5^0 \cdot (-1) \cdot \ln 5$$

$$= -\ln 5$$

$$= \ln\frac{1}{5}$$

23. $f(x) = x^2 + 3x - 5 \quad \Rightarrow f''(x) = ?$

A) 2   B) 3   C) 4   D) 5   E) 6

Solution:

$$f(x) = x^2 + 3x - 5$$

$$f'(x) = 2x + 3$$

$$f''(x) = 2$$

Correct Answer - A

QUESTIONS

1. $f:R \rightarrow R, f(x) = \sqrt{x^2 + 4x + 3}$

$$\Rightarrow \frac{df(x)}{dx} = f'(x) = ?$$

A) $\dfrac{2x + 4}{\sqrt{2x + 4}}$ B) $\dfrac{2x + 4}{\sqrt{x^2 + 4x + 3}}$ C) $\dfrac{x + 2}{\sqrt{x^2 + 4x + 3}}$

D) $(x + 2) \cdot \sqrt{x^2 + 4x + 3}$

E) $(x^2 + 4x + 3)\sqrt{2x + 4}$

Solution:

$$f(x) = \sqrt{x^2 + 4x + 3}$$

$$\frac{df(x)}{dx} = f'(x) = \frac{2x + 4}{2\sqrt{x^2 + 4x + 3}}$$

$$= \frac{2 \cdot (x + 2)}{2 \cdot \sqrt{x^2 + 4x + 3}}$$

$$= \frac{x + 2}{\sqrt{x^2 + 4x + 3}}$$

Correct Answer - C

2. $f(x) = \dfrac{ax^2 + b}{bx + a} \Rightarrow \dfrac{df(x)}{dx} = f'(x)$

$f'(0) = -4 \Rightarrow \dfrac{b^2}{a^2} = ?$

A) $\dfrac{1}{2}$ B) 1 C) 2 D) 3 E) 4

Solution:

$f(x) = \dfrac{ax^2 + b}{bx + a}$

$\dfrac{df(x)}{dx} = f'(x) = \dfrac{2ax \cdot (bx + a) - b \cdot (ax^2 + b)}{(bx + a)^2}$

$f'(0) = \dfrac{2 \cdot a \cdot 0 (b \cdot 0 + a) - b \cdot (a \cdot 0 + b)}{(b \cdot 0 + a)^2}$

$= \dfrac{-b^2}{a^2} = -4 \Rightarrow \dfrac{b^2}{a^2} = 4$

Correct Answer - E

3. $f(x) = ax^2 - bx \Rightarrow \dfrac{df(x)}{dx} = f'(x)$

$f'(0) = -3 \Rightarrow f'(b) = ?$

A) $2a - 3$ B) $3a - 2$ C) $2a + 3$

D) $3a + 2$ E) $6a - 3$

Solution:

$f(x) = ax^2 - bx$

$\dfrac{df(x)}{dx} = f'(x) = 2ax - b$

$f'(0) = 2 \cdot a \cdot 0 - b$

$\quad -b = -3$

$f'(b) = f'(3)$

$\qquad = 2 \cdot a \cdot 3 - 3$

$\qquad = 6a - 3$

Correct Answer - E

4. $f(x) = \dfrac{\sin x}{1 + \cos x}$ $\Rightarrow \dfrac{df(x)}{dx} = f'(x) = ?$

A) $\dfrac{1}{\cos x + 1}$ B) $\dfrac{\cos x}{1 + \sin x}$ C) $\dfrac{1}{\sin x}$

D) $\cos x$ E) $\sin x$

Solution:

$f(x) = \dfrac{\sin x}{1 + \cos x}$

$$\frac{df(x)}{dx} = f'(x) = \frac{\cos x(1+\cos x) - (-\sin x)\cdot\sin x}{(1+\cos x)^2}$$

$$= \frac{\cos x + \cos^2 x + \sin^2 x}{(1+\cos x)^2}$$

$$= \frac{\cos x + 1}{(1+\cos x)^2}$$

$$= \frac{1}{1+\cos x}$$

Correct Answer - A

5. $f(x) = 2\sin x - \cos x \Rightarrow \dfrac{d}{dx} f\!\left(\dfrac{\pi}{4}\right) = f'\!\left(\dfrac{\pi}{4}\right) = ?$

A) 2   B)$\dfrac{3}{2}$ C)$\dfrac{3\sqrt{2}}{2}$ D) $2\sqrt{2}$ E) $\sqrt{2}$

Solution:

$f(x) = 2\sin x - \cos x$

$f'(x) = 2\cdot\cos x - (-\sin x)$

$\quad = 2\cos x + \sin x$

$f'\!\left(\dfrac{\pi}{4}\right) = 2\cdot\cos\dfrac{\pi}{4} + \sin\dfrac{\pi}{4}$

$\quad = 2\cdot\dfrac{\sqrt{2}}{2} + \dfrac{\sqrt{2}}{2}$

$$= \frac{3\sqrt{2}}{2}$$

Correct Answer - C

6. $f(x) = (2x^3 + 3x^2)e^{-2x} \quad \Rightarrow \quad e^{2x}\dfrac{df(x)}{2dx} = ?$

A) $3(x^2 + x)$ B) $x^3 + x^2$ C) $3x - 2x^3$

D) $6 \cdot (x^2 + x) \cdot e^{-2x}$ E) $6 \cdot (x^2 + x) \cdot e^{2x}$

Solution:

$$f(x) = (2x^3 + 3x^2) \cdot e^{-2x}$$

$$\frac{df(x)}{2dx} = (6x^2 + 6x) \cdot e^{-2x} + e^{-2x} \cdot (-2) \cdot (2x^3 + 3x^2)$$

$$= (6x^2 + 6x) \cdot e^{-2x} - 2 \cdot (2x^3 + 3x^2) \cdot e^{-2x}$$

$$= e^{-2x} \cdot (6x^2 + 6x - 2) \cdot (2x^3 + 3x^2))$$

$$= e^{-2x} \cdot (6x^2 + 6x - 4x^3 - 6x^2)$$

$$= e^{-2x} \cdot (6x - 4x^3)$$

$$e^{2x} \cdot \frac{df(x)}{2dx} = \frac{e^{2x} \cdot e^{-2x} \cdot (6x - 4x^3)}{2}$$

$$= \frac{6x - 4x^3}{2}$$

$$= \frac{2(3x - 2x^3)}{2}$$

$$= 3x - 2x^3$$

7. $x = 3 \cdot t^2 + 6 \cdot t$, $y = 2 \cdot t^3 - 6 \cdot t$ $\Rightarrow \dfrac{dy}{dx} = ?$

A) t B) $t - 1$ C) $t + 1$ D) $\dfrac{t - 1}{t + 1}$ E) $\dfrac{t + 1}{t - 1}$

Solution:

$$\begin{cases} x = 3t^2 + 6 \cdot t \\ y = 2t^3 - 6 \cdot t \end{cases} \Rightarrow \frac{dy}{dx} = \frac{\dfrac{dy}{dt}}{\dfrac{dx}{dt}}$$

$$= \frac{6 \cdot t^2 - 6}{6 \cdot t + 6}$$

$$= \frac{6 \cdot (t^2 - 1)}{6 \cdot (t + 1)}$$

$$= \frac{(t - 1) \cdot (t + 1)}{t + 1}$$

$$= t - 1$$

8. $x = e^{3t} \cdot \cos t$, $y = e^{3t} \sin t$ $\Rightarrow \dfrac{dy}{dx}\left(\dfrac{\pi}{4}\right) = ?$

A) – 4   B) – 2   C) 2   D) 3   E) 4

Solution:

$$\begin{cases} x = e^{3t} \cdot \cos t \\ y = e^{3t} \cdot \sin t \end{cases} \Rightarrow \frac{dy}{dx} = \frac{\dfrac{dy}{dt}}{\dfrac{dx}{dt}}$$

$$= \frac{e^{3t} \cdot 3 \cdot \sin t + \cos t \cdot e^{3t}}{e^{3t} \cdot 3 \cdot \cos t - \sin t \cdot e^{3t}}$$

$$= \frac{e^{3t} \cdot (3 \sin t + \cos t)}{e^{3t} \cdot (3 \cos t - \sin t)}$$

$$\frac{dy}{dx}\left(\frac{\pi}{4}\right) = \frac{3 \cdot \sin\dfrac{\pi}{4} + \cos\dfrac{\pi}{4}}{3 \cdot \cos\dfrac{\pi}{4} - \sin\dfrac{\pi}{4}}$$

$$= \frac{3 \cdot \dfrac{\sqrt{2}}{2} + \dfrac{\sqrt{2}}{2}}{3 \cdot \dfrac{\sqrt{2}}{2} - \dfrac{\sqrt{2}}{2}} = \frac{\dfrac{4\sqrt{2}}{2}}{\dfrac{2\sqrt{2}}{2}} = 2$$

Correct Answer - C

9. $f(x) = x \cdot \sqrt{x^2 + 2x - 3} \Rightarrow \sqrt{5} f'(2) = ?$

A) – 3   B) – 2   C) 8   D) 10   E) 11

Solution:

$$f(x) = x \cdot \sqrt{x^2 + 2x - 3}$$

$$f'(x) = 1 \cdot \sqrt{x^2 + 2x - 3} + \frac{2x + 2}{2\sqrt{x^2 + 2x - 3}} \cdot x$$

$$f'(x) = \sqrt{x^2 + 2x - 3} + \frac{2(x + 1) \cdot x}{2\sqrt{x^2 + 2x - 3}}$$

$$f'(x) = \sqrt{x^2 + 2x - 3} + \frac{(x + 1) \cdot x}{\sqrt{x^2 + 2x - 3}}$$

$$f'(2) = \sqrt{2^2 + 2 \cdot 2 - 3} + \frac{(2 + 1) \cdot 2}{\sqrt{2^2 + 2 \cdot 2 - 3}}$$

$$f'(2) = \sqrt{5} + \frac{6}{\sqrt{5}}$$

$$\sqrt{5} \cdot f'(2) = \sqrt{5} \cdot \left(\sqrt{5} + \frac{6}{\sqrt{5}} \right) = 5 + 6 = 11$$

Correct Answer - E

10. $x = t + \dfrac{1}{t}$, $y = t^2 - \dfrac{2}{t}$, $t = 2 \Rightarrow \dfrac{dy}{dx} = ?$

A) 6   B) 4   C) 2   D) – 2   E) – 4

Solution:

$$\begin{cases} x = t + \dfrac{1}{t} \\ y = t^2 - \dfrac{2}{t} \end{cases} \Rightarrow \dfrac{dy}{dx} = \dfrac{\dfrac{dy}{dt}}{\dfrac{dx}{dt}}$$

$$= \dfrac{2 \cdot t + \dfrac{2}{t^2}}{1 - \dfrac{1}{t^2}}$$

$$= \dfrac{\dfrac{2 \cdot t^3 + 2}{t^2}}{\dfrac{t^2 - 1}{t^2}}$$

$$= \dfrac{2 \cdot t^3 + 2}{t^2 - 1}$$

$$t = 2 \Rightarrow \dfrac{dy}{dx} = \dfrac{2 \cdot 2^3 + 2}{2^2 - 1} = \dfrac{18}{3} = 6$$

Correct Answer - A

11. $\dfrac{1}{\sin 2x} \cdot \dfrac{d}{dx}(\sin^2 x) = ?$

$A) \sin x$ $B) \cos x$ $C) \cos 2x$ $D) -1$ $E) 1$

Solution:

$$\frac{1}{\sin 2x} \cdot \frac{d}{dx}(\sin^2 x) = \frac{1}{\sin 2x} \cdot 2 \cdot \sin x \cdot \cos x$$

$$= \frac{1}{\sin 2x} \cdot \sin 2x$$

$$= 1$$

Correct Answer - E

12. $x = 2t - \dfrac{1}{2}$, $y = t^2 + 2$ $\Rightarrow \dfrac{d^2 y}{dx^2} = ?$

A) 1   B)$\dfrac{1}{2}$ C)$\dfrac{1}{3}$ D)$\dfrac{3}{4}$ E)$\dfrac{7}{5}$

Solution:

$$\left. \begin{matrix} x = 2 \cdot t - \dfrac{1}{2} \\ y = t^2 + 2 \end{matrix} \right\} \Rightarrow \frac{d^2 y}{dx^2} = \frac{d}{dx}\left(\frac{dy}{dx}\right)$$

$$= \frac{d}{dx}\left(\frac{\frac{dy}{dt}}{\frac{dx}{dt}}\right)$$

$$= \frac{d}{dx}\left(\frac{2 \cdot t}{2}\right)$$

$$= \frac{d}{dx}(t) = \frac{1}{2}$$

13. $y = e^{2t}$, $x = \cos e^{2t}$ $\Rightarrow \dfrac{dy}{dx} = ?$

     A) $\cos 2x$ B) $\sin 2x$ C) $\sin x$

   D) $\dfrac{1}{\sin(\arcsin x)}$ E) $-\dfrac{1}{\sin(\arccos x)}$

Solution:

$$\frac{dy}{dx} = \frac{\dfrac{dy}{dt}}{\dfrac{dx}{dt}} = \frac{2e^{2\cdot t}}{-2e^{2\cdot t}\sin(e^{2\cdot t})}$$

$$= \frac{-1}{\sin(e^{2\cdot t})} = \frac{-1}{\sin(\arccos x)}$$

14. $f(x) = x \cos x$

$$\frac{d^2 f(x)}{dx^2}\bigg|_{x = \frac{\pi}{2}}$$

     A) – 3   B) – 2   C) – 1   D) 0   E) 1

Solution:

$$\frac{d}{dx}\left(\frac{df(x)}{dx}\right) = \frac{d}{dx}(\cos x - \sin x \cdot x)$$

$$= -\sin x - (\cos x \cdot x + \sin x)$$

$$= -\sin x - x \cdot \cos x - \sin x$$

$$= -2\sin x - x \cdot \cos x$$

$$x = \frac{\pi}{2} \Rightarrow -2 \cdot \sin\frac{\pi}{2} - \frac{\pi}{2} \cdot \cos\frac{\pi}{2} = -2 \cdot 1 - \frac{\pi}{2} \cdot 0$$

$$= -2$$

Correct Answer - B

15. $f(x) = \dfrac{x^2 - 5x + 6}{x^2 + 5x + 6} \Rightarrow \dfrac{df}{dx}(0) = f'(0) = ?$

A) $-\dfrac{5}{3}$ B) $-\dfrac{2}{3}$ C) $-\dfrac{1}{3}$ D) -2 E) -1

Solution:

$$\frac{df}{dx} = f'(x)$$

$$= \frac{(2x-5)\cdot(x^2 + 5x + 6) - (x^2 - 5x + 6)(2x + 5)}{(x^2 + 5x + 6)^2}$$

$$f'(0) = \frac{(-5)\cdot 6 - 6\cdot 5}{6^2} = \frac{-60}{36} = \frac{-5}{3}$$

16. $y = x \cdot e^{-x} + \ln 2 \quad \Rightarrow \dfrac{dy}{dx} = ?$

A) $x \cdot e^{-x}$ B) $e^{-x} - x \cdot e^{-x}$ C) $e^{-x} - x \cdot e^x + \dfrac{1}{2}$

D) $e^{-x} + \dfrac{1}{2}$ E) $x \cdot e^{-x} + 2$

Solution:

$\dfrac{dy}{dx} = e^{-x} - x \cdot e^{-x}$

17. $f(x) = \sqrt{x}(x^3 - 3) \Rightarrow \dfrac{df}{dx}(1) = f'(1) = ?$

A) -3 B) -1 C) 0 D) 1 E) 2

Solution:

$\dfrac{df}{dx} = f'(x) = \dfrac{1}{2\sqrt{x}} \cdot (x^3 - 3) + \sqrt{x} \cdot 3x^2$

$f'(1) = -1 + 3 = 2$

18. $f(x) = \ln(\cos x) \Rightarrow \dfrac{df}{dx}\left(\dfrac{\pi}{4}\right) = f'\left(\dfrac{\pi}{4}\right) = ?$

A) 0   B) – 1   C) 1   D) – e   E) e

Solution:

$$\frac{df}{dx} = f'(x) = \frac{1}{\cos x}\cdot(\cos x)' = \frac{-\sin x}{\cos x} = -\tan x$$

$$f'\left(\frac{\pi}{4}\right) = -\tan\left(\frac{\pi}{4}\right) = -1$$

Correct Answer - B

19. $f(x) = \ln\left(\sqrt{x^3 - 4}\right)\left(f'(x) = \dfrac{df(x)}{dx}\right) \Rightarrow f'(2) = ?$

A)$\dfrac{3}{2}$ B)$\dfrac{1}{2}$ C) 0   D) 1   E) 2

Solution:

$$f'(x) = \frac{\dfrac{3x^2}{2\sqrt{x^3-4}}}{\sqrt{x^3-4}}$$

$$f'(2) = \frac{\dfrac{12}{4}}{2} = \frac{3}{2}$$

Correct Answer - A

20. $f(x) = (x-2)^4$, $\left(f''(x) = \dfrac{d^2f(x)}{dx^2}\right) \Rightarrow f''(3) = ?$

A) 4   B) 6   C) 8   D) 10   E) 12

Solution:

$f'(x) = 4(x-2)^3$

$f''(x) = 12(x-2)^2$

$f''(3) = 12(3-2)^2 = 12$

Correct Answer - E

21. $f(x) = x \cdot |9 - x^2| \Rightarrow f'(4) = ?$

A) 39   B) 28   C) 7   D) −7   E) −39

Solution:

$f(x) = x \cdot |9 - x^2|$

$$\Rightarrow f'(x) = 1 \cdot |9 - x^2| + x \cdot \frac{(-2x) \cdot |9 - x^2|}{9 - x^2}$$

$$= 7 + \frac{4 \cdot (-8) \cdot 7}{-7} = 39$$

22. $f(x) = \ln(\cos e^x) \Rightarrow f'\left(\ln\dfrac{\pi}{4}\right) = ?$

A)$\dfrac{-\pi}{4}$ B)$\dfrac{-\pi}{2}$ C)$\dfrac{\pi}{4}$ D) $e^{\pi/4}$ E) $e^{\pi/2}$

Solution:

$f(x) = \ln(\cos e^x)$

$\Rightarrow f'(x) = \dfrac{-\sin e^x \cdot (e^x)'}{\cos e^x}$

$= (-\tan e^x) \cdot (e^x)$

$= \left(-\tan e^{\left(\ln\frac{\pi}{4}\right)}\right) \cdot (e^{\left(\ln\frac{\pi}{4}\right)})$

$= -\left(\tan\dfrac{\pi}{4}\right)\cdot\dfrac{\pi}{4}$

$= -\dfrac{\pi}{4}$

Correct Answer - A

50

Test 1

1. $f(x) = x^2 + \sqrt{x} \implies f'(9) = ?$

A) 18   B)$\dfrac{109}{6}$ C)$\dfrac{55}{3}$ D)$\dfrac{37}{2}$ E)$\dfrac{56}{3}$

2. $f(x) = \dfrac{x}{x^2 - 1} \implies f'(2) = ?$

A) $-\dfrac{7}{9}$ B)$\dfrac{2}{3}$ C) $-\dfrac{5}{9}$ D) $-\dfrac{4}{9}$ E) $-\dfrac{1}{3}$

3. $\begin{cases} x = \ln t \\ y = t^2 \end{cases} \implies \dfrac{d^2y}{dx^2} = ?$

A) t B) t^2 C) $2t^2$ D) $4t^2$ E) $8t^2$

4. $f(x) = \ln(\cos x + \sin x) \implies f'(x) = ?$

$$A)\frac{\cos 2x}{\sin x + \cos x} \qquad B)\frac{\sin x \cdot \cos x}{\sin x + \cos x} \qquad C)\frac{\sin x - \cos x}{\cos x + \sin x}$$

$$D)\frac{\cos x - 2\sin x}{\cos x + \sin x} \qquad E)\frac{\cos x - \sin x}{\cos x + \sin x}$$

5. $x^2 - xy - y^2 + 5 = 0 \quad \Rightarrow \frac{dy}{dx} = ?$

$$A)\frac{y - x}{x + 2y} \qquad B)\frac{y - 2x}{x + 2y} \qquad C)\frac{y + 2x}{x - 2y}$$

$$D)\frac{2x - y}{x + 2y} \qquad E)\frac{y + x}{x - 2y}$$

6. $f(x) = (3 - 2x)^6 \quad \Rightarrow f'(1) = ?$

A) -12 B) -6 C) -3 D) 3   E) 6

7. $y = \sin^3 x \Rightarrow \frac{dy}{dx} = ?$

A) $3\sin^3 x \cdot \cos x$ B) $3\sin^2 x \cdot \cos x$

C) $3\sin x \cdot \cos x$ D) $3\sin^2 x \cdot \tan x$

E) $3\sin x \cdot \cot x$

8. $f(x) = \cos (\sin x) \quad \Rightarrow f'\left(\frac{\pi}{2}\right) = ?$

A) $-\dfrac{\sqrt{3}}{2}$ B) $-\dfrac{1}{2}$ C) 0   D)$\dfrac{1}{2}$ E)$\dfrac{\sqrt{3}}{2}$

9. $f(x) = \ln\dfrac{x-a}{x+a}$ $\Rightarrow$ $f'(x) = ?$

A) $\dfrac{2a+2x}{x^2-a^2}$ B) $\dfrac{2x^2}{a^2-x^2}$ C) $\dfrac{2x^2}{x^2-a^2}$

D) $\dfrac{2a}{a^2-x^2}$ E) $\dfrac{2a}{x^2-a^2}$

10. $\dfrac{1}{2\cdot\cos 2x}\cdot\dfrac{d}{dx}(\sin^2 x) = ?$

A) $\dfrac{\cot 2x}{\tan x}$ B) $\dfrac{\cot x}{2}$ C) $\dfrac{\cot 2x}{2}$

D) $\dfrac{\tan 2x}{2}$ E) $\dfrac{\tan x}{2}$

11. $f(x) = e^{\sin x}$ $\Rightarrow$ $f'(0) = ?$

A) $-e^2$ B) $-e$ C) 1   D) e E) e^2

12. $f(x) = 3^{\cos x}$ $\Rightarrow$ $f'\!\left(\dfrac{\pi}{2}\right) = ?$

A) $-\ln 6$ B) $-\ln 3$ C)$\ln 2$ D)$\ln 3$ E)$\ln 6$

13. $\sin(xy) = x^2 + y^3 \Rightarrow \dfrac{dy}{dx} = ?$

A) $\dfrac{2x - y \cdot \cos(xy)}{x \cdot \cos(xy) - 3y^2}$

B) $\dfrac{2x + y \cdot \sin(xy)}{x \cdot \cos(xy) - 3y^2}$

C) $\dfrac{2x + y \cdot \cos(xy)}{x \cdot \sin(xy) - 3y^2}$

D) $\dfrac{2x - y \cdot \cos(xy)}{x \cdot \cos(xy) + 3y^2}$

E) $\dfrac{2x + y \cdot \cos(xy)}{x \cdot \cos(xy) + 3y^2}$

14. $\begin{cases} x = 2 \cdot \sin t \\ y = 3 \cdot \cos t \end{cases} \Rightarrow \dfrac{d^2 y}{dx^2} = ?$

A) $\dfrac{-3}{4\cos^3 t}$

B) $\dfrac{-3}{4\sin t}$

C) $\dfrac{3}{4\tan t}$

D) $\dfrac{3}{4\sin t}$

E) $\dfrac{3\sin t}{4\cos t}$

15. $f(x) = 5^x - 8^x \Rightarrow f'(0) = ?$

A) $\ln\dfrac{3}{8}$

B) $\ln\dfrac{1}{2}$

C) $\ln\dfrac{5}{8}$

D) $\ln\dfrac{3}{4}$

E) $\ln\dfrac{7}{8}$

16. $f(x) = \sqrt[5]{-x^3 + 2x} \Rightarrow f'(-1) = ?$

54

A) – 5   B) $-\dfrac{1}{5}$ C) 1   D)$\dfrac{1}{5}$ E) 5

17. $f(x) = \ln(\cos x)$ $\Rightarrow$ $f'\left(\dfrac{\pi}{3}\right) = ?$

A) $-2\sqrt{3}$ B) $-\sqrt{3}$ C) -1 D) $\sqrt{3}$ E) $2\sqrt{3}$

18. $y = \log_2 x^2$ $\Rightarrow \dfrac{dy}{dx} = ?$

A) $\dfrac{2+x}{\ln 2}$ B) $\dfrac{2}{x \cdot \ln 2}$ C) $\dfrac{1}{x\ln 2}$

D) $\dfrac{x \cdot \ln 2}{x + \ln 2}$ E) $\dfrac{\ln 2}{x + \ln 2}$

19. $y = (1 - x^2)^3$ $\Rightarrow \dfrac{d^2 y}{dx^2}\bigg|_{x=1} = ?$

A) -36 B) -24 C) -12 D) 0   E) 24

20. $f(x) = \sqrt{1 + x^3}$ $\Rightarrow$ $f'(2) = ?$

A) 2   B) 3   C) 4   D) 5   E) 6

21. $x = f(t) = t^3 - 1$

$$y = g(t) = 2t^2 + 2t$$

$$\Rightarrow \frac{dy}{dx}\bigg|_{t=1} = \ ?$$

A) 1   B) 2   C) 3   D) 4   E) 5

Answers					
1. B	2. C	3. D	4. E	5. D	6. A
7. B	8. C	9. E	10. D	11. C	12. B
13. A	14. A	15. C	16. B	17. B	18. B
19. D	20. A	21. B			

Chapter 16   The Derivative

Test 2

1. $f(x) = 2x^3 - 3x^2 - 12x + 20 \ \Rightarrow \ f'(-1) + f''(1) = \ ?$

A) – 6   B) – 1   C) 0   D) 1   E) 6

2. $f(x) = x^3 + ax^2 + bx + 3$,

$f'(1) = \ -2, f'(2) = 0 \ \Rightarrow b = \ ?$

A) 2   B) 3   C) 5   D) 6   E) 7

3. $y = \dfrac{2x+1}{x-2} \ \Rightarrow \dfrac{dy}{dx}\bigg|_{x=3} = \ ?$

A) – 5   B) – 4   C) – 3   D) 3   E) 4

4. $f(x) = (2x - 1)^3 \cdot \ln x \Rightarrow f'(1) = ?$

A) – 2   B) – 1   C) 0   D) 1   E) 2

5. $f(x) = x \cdot e^x \Rightarrow f'(2) + f''(2) = ?$

A) $6e^2$ B) $7e^2$ C)$\dfrac{1}{6e^2}$ D)$\dfrac{1}{7e^2}$ E)$\dfrac{6}{e^2}$

6. $f(x) = \sqrt{e^x} \cdot \ln(x^2) \Rightarrow f'(1) = ?$

A)$\dfrac{2}{\sqrt{e}}$ B)$\dfrac{4}{\sqrt{e}}$ C) $2\sqrt{e}$ D) $3\sqrt{e}$ E) $4\sqrt{e}$

7. $f(3x - 2) = x^3 - 3x + 1 \Rightarrow f'(4) = ?$

A) 2   B) 3   C) 4   D) 5   E) 6

8. $\begin{cases} f(x) = \sqrt{x} \\ g(x) = x^2 + 3 \end{cases} \Rightarrow (f \circ g)'(1) = ?$

A)$\dfrac{1}{3}$ B)$\dfrac{1}{2}$ C) 1   D)$\dfrac{3}{2}$ E) 2

9. $f(x) = \cos(2x)$

$$\Rightarrow \lim_{h \to 0} \frac{f\left(\frac{\pi}{6} + h\right) - f\left(\frac{\pi}{6}\right)}{h} = ?$$

A) $-\sqrt{3}$ B) $-\dfrac{1}{\sqrt{3}}$ C) 1   D) 2   E) $\sqrt{3}$

10. $f(x) = \arccos(2x + 1)$ $\Rightarrow f'\left(-\dfrac{1}{2}\right) = ?$

A) -2 B) -1 C) 0   D) $\dfrac{1}{2}$ E) $-\dfrac{1}{2}$

11. $f(x) = \tan(\sqrt[3]{x})$ $\Rightarrow f'(\pi^3) = ?$

A) π^2 B) $2\pi^2$ C) $3\pi^2$ D) $\dfrac{1}{3\pi^2}$ E) $\dfrac{1}{2\pi^2}$

12. $f(x) = \sin^2 \sqrt{x}$ $\Rightarrow \dfrac{df(x)}{dx} = f'(x) = ?$

A) $\dfrac{1}{2\sqrt{x}} \sin 2\sqrt{x}$ B) $\dfrac{1}{\sqrt{x}} \sin 2\sqrt{x}$ C) $\dfrac{1}{2\sqrt{x}} \sin\sqrt{x}$

D) $\dfrac{1}{2\sqrt{x}} \sin 2\sqrt{x} \cdot \cos \sqrt{x}$ E) $\dfrac{1}{4\sqrt{x}} \sin 2\sqrt{x}$

13. $f(x) = \ln(\arctan x)$ $\Rightarrow f'(1) = ?$

58

A)$\dfrac{\pi}{3}$ B)$\dfrac{2}{\pi}$ C)$\dfrac{\pi}{4}$ D)$\dfrac{1}{2\pi}$ E) π

14. $f(x) = \log x^3$ $\Rightarrow f'\left(\dfrac{1}{\ln 10}\right) = ?$

A) 3   B) 1   C) 0   D) $3\ln 10$ E) $(\ln 10)^2$

15. $f(x) = 5^{3x-3}$, $f'(a) = \ln 5^{375}$ $\Rightarrow a = ?$

A) 4   B) 3   C) 0   D) 1   E) 2

16. $\begin{cases} x = t^3 - 2t^2 + 3t \\ y = t^3 - 2t \end{cases}$ $\Rightarrow \dfrac{dy}{dx}\bigg|_{t=1} = ?$

A)$\dfrac{1}{3}$ B)$\dfrac{1}{2}$ C) 1   D) 2   E) 3

17. $\begin{cases} y = e^x \\ x = \cos t \end{cases}$ $\Rightarrow \dfrac{dy}{dt}\bigg|_{t=\frac{\pi}{3}} = ?$

A) $-\dfrac{\sqrt{3e}}{2}$ B) $-\dfrac{\sqrt{e}}{4}$ C) $-\dfrac{\sqrt{2e}}{3}$ D) $\sqrt{3e}$ E) $\sqrt{2e}$

18. $\begin{cases} y = e^t + 2t \\ x = e^{2t} \end{cases}$ $\Rightarrow \dfrac{d^2y}{dx^2}\bigg|_{t=0} = ?$

A) $-\dfrac{3}{4}$ B) $-\dfrac{5}{4}$ C) $-\dfrac{7}{4}$ D) $\dfrac{1}{2}$ E) $\dfrac{3}{2}$

19. $2x^2 + 3xy - 4y^2 = 0 \quad \Rightarrow \dfrac{dy}{dx} = ?$

A) $\dfrac{4x - 3y}{3x + 8y}$ B) $\dfrac{-4x - 3y}{3x - 8y}$ C) $\dfrac{4x + 3y}{3x + 8y}$

D) $\dfrac{2x - 3y}{3x + 4y}$ E) $\dfrac{2x + 3y}{3x - 8y}$

20. $f(x) = (x^2)^{\sin x} \quad \Rightarrow \quad f'\left(\dfrac{\pi}{2}\right) = ?$

A) $\dfrac{2}{\pi}$ B) $\dfrac{4}{\pi}$ C) π D) 2π E) 3π

21. $f(x) = x(2 - \ln x) \quad \Rightarrow \dfrac{df(e)}{dx} = ?$

A) 0   B) 2   C) 3   D) 4   E) 5

22. $f(x) = \dfrac{\ln x}{x^2 + 1} \quad \Rightarrow \dfrac{df(1)}{dx} = f'(1) = ?$

A) 1   B) $\dfrac{1}{2}$ C) $\dfrac{3}{1}$ D) $\dfrac{1}{4}$ E) $\dfrac{3}{4}$

Chapter 16   **The Derivative**

Test 3

1. $f(x) = ax^2 - 3x + 1$

   $\dfrac{d}{dx} f(1) = 5 \quad \Rightarrow a = ?$

   A) 6   B) 5   C) 4   D) 3   E) 2

2. $f(x) = x^4 - 3x^2 + 6x + 3$

$$\Rightarrow \lim_{x \to 1} \frac{f(x) - f(1)}{x - 1} = ?$$

A) 2   B) 4   C) 5   D) 10   E) 15

3. $f(x) = ax^2 - bx$

$$\Rightarrow \frac{d}{dx} f(1) = ?$$

A) $6a - b$ B) $3a + 2$ C) $2a + b$

D) $3a - 2b$ E) $2a - b$

4. $a < 0$,

$$f(x) = \frac{1}{3}x^3 + x^2 - 3x + 7$$

$$\frac{d}{dx} f(a) = 0 \quad \Rightarrow a = ?$$

A) -4 B) -3 C) -2 D) 1   E) 6

5. $f(x) = x^3 - bx^2 + 6x + 3$

$$\frac{d}{dx} f(2) = \frac{d}{dx} f(1)$$

$$\Rightarrow b = ?$$

$$A)\frac{7}{2} \quad B)\frac{9}{2} \quad C)\frac{11}{2} \quad D)\frac{15}{2} \quad E)\frac{19}{2}$$

6. $\begin{cases} f(x) = x^3 - ax^2 + bx + 3, \\ f(1) = 12; \ f'(2) = 8 \end{cases} \Rightarrow b = ?$

A) 4   B) 6   C) 8   D) 12   E) 16

7. $\forall x \in N^+, f(0) = 0$

   $f(x) - f(x - 1) = x + 1$

   $\Rightarrow f'(6) = ?$

$$A)\frac{7}{2} \quad B)\frac{9}{2} \quad C)\frac{11}{2} \quad D)\frac{13}{2} \quad E)\frac{15}{2}$$

8. $\forall x \in Z^+, f(0) = 0$

   $f(x) = x^2 + f(x - 1)$

   $\Rightarrow f'(1) + f'(0) = ?$

$$A)\frac{7}{3} \quad B)\frac{4}{3} \quad C) 1 \quad D)\frac{8}{3} \quad E) 3$$

9. $\forall f'(1) + f'(0) \in R, f(x) = f(-x)$

   $f(x) = 4x^4 - 2ax^2 + 4 - f(-x)$

63

$\frac{d}{dx}f(2) = 32 \implies a = ?$

A) 2   B) 4   C) 5   D) 6   E) 8

10. $\forall x \in R, f(x) + f(-x) = 0$

$2f(x) = 5x^3 - 10ax - 20 + 3f(-x)$

$\frac{d}{dx}f(3) = 15 \implies a = ?$

A) 2   B) 3   C) 4   D) 5   E) 6

11. $f(2x - 1) = 2x^2 - 6x + 4$

$\implies \frac{df(x)}{dx} = ?$

A) $2x + 1$ B) $2x - 1$ C) $x - 2$

D) $2x - 2$ E) $x - 3$

12. $f(2x - 1) = x^3 - x^2 + 4x + 1$

$\implies f(3) + f'(3) = ?$

A) 15   B) 17   C) 19   D) 21   E) 23

13. $f(3x - 4) = x^3 - 6x^2 + 7$

64

$$\Rightarrow f(2) + f'(2) = ?$$

A) -15 B) -14 C) -13 D) -12 E) -10

14. $f(x) = 2x - 3$

$(gof)(x) = 4x^2 - 8x + 7$

$$\Rightarrow \frac{dg(x)}{dx} = ?$$

A) $x + 2$ B) $2(x + 1)$ C) $2(x + 2)$

D) $x - 2$ E) $2(x - 2)$

15. $f(x^3 + 2) = 3x^9 - 6x^6 + 5$

$$\Rightarrow \frac{df(x)}{dx} = ?$$

A) $9x^2 - 32x$ B) $9x^2 - 36x + 40$ C) $9x^2 - 48x + 60$

D) $9x^2 - 42x + 50$ E) $9x^2 - 18x - 70$

16. $f(x) = (ax^2 - 1)(x^2 + 2x + 3)$

$$\frac{d}{dx}f(2) = 130$$

$$\Rightarrow a = ?$$

A) 2   B) 3   C) 5   D) 13   E) 21

17. $g(x) = x^2 \cdot f(x)$

   $g'(4) = -48$

   $\Rightarrow a = ?$

A) 2   B) 3   C) 4   D) 5   E) 6

18. $f(x) = (x^2 + 1) \cdot g(2x + 1)$

   $g(5) = 6, \ g'(5) = 3$

   $\Rightarrow f'(2) = ?$

A) 36   B) 46   C) 54   D) 56   E) 58

19. $y = \dfrac{x}{(x-1)^2}$

   $\Rightarrow (x-1)^4 \cdot \dfrac{dy}{dx} = ?$

A) $1 - x^2$ B) $1 + x^2$ C) $x^2 - 1$

D) $(x^2 - 1)^2$ E) $(x-1)^3$

20. $f(x) = \dfrac{ax^2 + 1}{x^2 + 1}$

$$\frac{d}{dx} f(2) = 4 \quad \Rightarrow a = \,?$$

A) 15   B) 21   C) 26   D) 29   E) 33

Answers					
1. C	2. B	3. E	4. B	5. B	6. D
7. E	8. A	9. E	10. E	11. C	12. C
13. C	14. B	15. C	16. A	17. B	18. C
19. A	20. C				

Test 4

1. $y = \sqrt[7]{x^2}$

$\Rightarrow \sqrt[7]{x^5} \cdot \dfrac{dy}{dx} = ?$

A) $\dfrac{1}{7}$ B) $\dfrac{2}{7}$ C) $\dfrac{1}{\sqrt{x}}$ D) $\dfrac{3}{7}$ E) $\dfrac{x}{\sqrt{x}}$

2. $f(x) = \sqrt[3]{x^2} \ (a \neq 0)$

$\dfrac{df(a)}{dx} = f(a) \Rightarrow a = ?$

A) 4   B) 2   C) $\dfrac{2}{3}$ D) $\dfrac{1}{2}$ E) -1

3. $a < 0$

$f(x) = \sqrt[3]{x^2 + a}$

$f'(1) = \dfrac{1}{6} \Rightarrow a = ?$

A) -9 B) -7 C) -5 D) -3 E) -1

4. $f(x) = \sqrt[3]{x + a}$

$f^{-1}(1) + (f^{-1})'(1) = 2 \Rightarrow a = ?$

A) 1   B) 2   C) 3   D) 4   E) 5

5. $f(x) = \dfrac{(x-2)^3}{2}$

$\Rightarrow (f^{-1})(4) + (f^{-1})'(4) = ?$

A)$\dfrac{16}{17}$ B)$\dfrac{1}{25}$ C)$\dfrac{5}{3}$ D)$\dfrac{25}{6}$ E) 25

6. $f(x) = \dfrac{(x-1)^3}{4}$

$(f^{-1})(16)\cdot(f^{-1})'(16) = ?$

A)$\dfrac{5}{4}$ B)$\dfrac{5}{8}$ C)$\dfrac{5}{12}$ D)$\dfrac{5}{16}$ E)$\dfrac{5}{48}$

7. $y = \dfrac{x}{3\sqrt{x}}$ $\Rightarrow \dfrac{dy}{dx} = ?$

A)$\dfrac{1}{3\sqrt{x}}$ B)$\dfrac{1}{2\sqrt{x}}$ C)$\dfrac{1}{4\sqrt{x}}$ D)$\dfrac{1}{6\sqrt{x}}$ E)$\dfrac{1}{9\sqrt{x}}$

8. $f(x) = \sqrt[3]{\dfrac{x}{16}}$

$\Rightarrow f(2)\cdot(f^{-1})'(1) = ?$

A) 12   B) 18   C) 20   D) 24   E) 28

9. $f(x) = (x+1)^2 \cdot (3x+1)^2$

$$\Rightarrow \frac{df(1)}{dx} = ?$$

A) 96   B) 120   C) 144   D) 156   E) 160

10. $f(x) = \dfrac{\sqrt{x}}{1+\sqrt{x}} \quad \Rightarrow \dfrac{df(4)}{dx} = ?$

A) 12   B) 1   C)$\dfrac{1}{36}$ D)$\dfrac{1}{16}$ E)$\dfrac{1}{12}$

11. $y = (x^2+1)\sqrt{x} \quad \Rightarrow \dfrac{dy}{dx} = y' = ?$

A) $\dfrac{x^2+1}{2\sqrt{x}}$ B) $\dfrac{2x^2+1}{2\sqrt{x}}$ C) $\dfrac{5x^2+1}{2\sqrt{x}}$

D) $\dfrac{3x^2+2}{2\sqrt{x}}$ E) $\dfrac{6x^2+3}{2\sqrt{x}}$

12. $f(x) = \log_2 x$

$$\Rightarrow \frac{d}{dx}f(1) = ?$$

A)$\log e$ B) $2\log e$ C)$\log_2 e$

D)$\dfrac{2}{\log e}$ E)$\dfrac{2}{10}\log e$

70

13. $a > 0, b > 0$

$f(x) = \ln(ax + 6)$

$\dfrac{d}{dx} f(1) = \dfrac{a}{5} \quad \Rightarrow a = ?$

A) – 2   B) – 1   C) 0   D) 1   E) 2

14. $a > 2$

$f(x) = a \cdot \ln(ax - 3)^2$

$\dfrac{d}{dx} f(2) = 8 \quad \Rightarrow a = ?$

A) 1   B) 3   C) 4   D) 5   E) 6

15. $f(x) = \log_a(x^2 + 3x + 1)$

$\dfrac{d}{dx} f(1) = \dfrac{1}{6} \quad \Rightarrow \quad a = ?$

A) e^3 B) e^4 C) e^5 D) e^6 E) e^{30}

16. $f(x) = x \cdot \ln x$

$\Rightarrow \dfrac{d}{dx} f(e) = ?$

A)$\dfrac{1}{2}$ B) 1   C) 2   D)$\dfrac{3}{2}$ E) 3

17. $f(x) = x^2 \cdot \ln(x^2 + 1)$

$\Rightarrow \dfrac{d}{dx} f(1) = ?$

A)$\ln 2e$ B)$\ln \dfrac{2}{e}$ C)$\ln 4e$ D)$\ln \dfrac{4}{e}$ E)$\ln 3e$

18. $y = \ln\left(\dfrac{x^2}{x^2 + 1}\right)$

$\Rightarrow (x^2 + 1) \cdot \dfrac{dy}{dx} = ?$

A) $2x^2 + 1$ B) $2x^2 + 2$ C)$\dfrac{2}{x^2}$

D)$\dfrac{2}{x}$ E) $4x^2 + 2$

19. $x > 0, \ f(x) = \ln \sqrt[k]{x^2}$

$\dfrac{df(4)}{dx} = \dfrac{1}{10} \quad \Rightarrow k = ?$

A) 5   B) 8   C) 10   D) 12   E) 18

1. B	2. C	3. A	4. B	5. D	6. C
7. D	8. D	9. E	10. C	11. C	12. A
13. B	14. E	15. D	16. C	17. C	18. D
19. A					

Chapter 16   The Derivative

Test 5

1. $f(x) = (3x^2 - 2x + 1)^4$ $\Rightarrow$ $f'(-1) = ?$

A) – 6912   B) – 6180   C) – 5900   D) 2300   E) 1800

2. $f(x) = x\sqrt{x} - x\sqrt{1-x}$ $\Rightarrow$ $f'\left(\dfrac{3}{4}\right) = ?$

A) $\dfrac{3\sqrt{3}+1}{4}$ B) $\dfrac{3\sqrt{3}+4}{4}$ C) $\dfrac{3\sqrt{3}+2}{4}$

D) $3\sqrt{3}-1$ E) $\dfrac{5\sqrt{3}+1}{4}$

3. $f(x) = 3\sin(3x + 5)$ $\Rightarrow$ $f'(7°) = ?$

A) $3\cos 26$ B) $9\cos 26$ C) $15\cos 26$

D) $3\sin 26$ E) $-\sin 26$

4. $f(x) = \dfrac{x \cdot \sin x}{1 + \tan x} \Rightarrow f'\left(\dfrac{\pi}{4}\right) = ?$

A) $\sqrt{2}$ B)$\dfrac{\sqrt{2}}{2}$ C)$\dfrac{\sqrt{2}}{4}$ D) $-\sqrt{2}$ E) -2

5. $f(x) = \arccos\sqrt{1 - 4x} \Rightarrow f'(x) = ?$

A) $-2\sqrt{4x^2}$ B) $-\sqrt{x + 4x^2}$ C)$\dfrac{-2}{\sqrt{x - 4x^2}}$

D)$\dfrac{1}{\sqrt{x - 4x^2}}$ E) $-\sqrt{x + 4x^2}$

6. $f(x) = \dfrac{2}{3}\arctan x + \dfrac{1}{3}\arctan\dfrac{x}{1 - x^2} \Rightarrow f'(1) = ?$

A) -1 B) 0 C)$\dfrac{1}{2}$ D)$\dfrac{3}{4}$ E) 1

7. $f(x) = (x + 1)^{1/x} \Rightarrow f'(1) = ?$

A)$\ln\dfrac{e}{4}$ B)$\ln 2e$ C) $2 - \ln 2$ D) 1 E) -1

8. $y = f(x)$, $\dfrac{x - y}{x - 2y} = 2 \Rightarrow f'(x) = ?$

74

A) $\dfrac{2}{3}$ B) 1   C) $\dfrac{1}{3}$ D) $-\dfrac{1}{3}$ E) $-\dfrac{4}{3}$

9. $f(3x) = x^2 \cdot g(x-2)$, $f'(3) = 2$, $g(-1) = 4$

$\Rightarrow g'(-1) = ?$

A) 3   B) 2   C) – 2   D) – 3   E) – 4

10. $f(x) = e^{\sin x} + \cos^2 x \quad \Rightarrow f'\left(\dfrac{\pi}{2}\right) = ?$

A) – 1   B) 0   C) 1   D) 2   E) e

11. $f(x) = \sin x$, $g(x) = 2^x \quad \Rightarrow (f \circ g)'(2) = ?$

A)$\ln 2 \cdot \cos 16$ B)$\ln 4 \cdot \cos 4$ C)$\ln 8 \cdot \cos 4$

D)$\ln 16 \cdot \cos 4$ E)$\ln 16 \cdot \cos 16$

12. $\begin{cases} x = e^t + t^2 \\ y = t \cdot e^t \end{cases} \Rightarrow \left.\dfrac{d^2y}{dx^2}\right|_{t=1} = ?$

A) $\dfrac{1}{e^2}$ B) $\dfrac{1}{2+e}$ C) $\dfrac{1}{4+e}$

D) $\dfrac{e}{(e+2)^2}$ E) $\dfrac{e}{(e+2)^3}$

13. $\begin{cases} x = t^2 - 2t \\ y = t^3 + t \end{cases} \Rightarrow \dfrac{d^2y}{dx^2}\bigg|_{t=1} = ?$

A) 2   B) $\dfrac{4}{3}$ C) 1   D) $\dfrac{3}{4}$ E) $\dfrac{1}{4}$

14. $\begin{cases} x = \ln t \\ y = \sin t \end{cases} \Rightarrow \dfrac{dy}{dx}\bigg|_{t=\pi} = ?$

A) $-\pi$ B) -2π C) 0   D) 1   E) 2

15. $f(x) = x \cdot \sin x \Rightarrow \dfrac{d^2 f(x)}{dx^2}\bigg|_{x=\frac{\pi}{2}} = ?$

A) -1 B) 0   C) 1   D) $\dfrac{\pi}{2}$ E) $-\dfrac{\pi}{2}$

16. $f(x) = x \cdot e^x \Rightarrow f''(x) - f'(x) = ?$

A) 0   B) x C) e^x D) xe^x E) $x + e^x$

17. $f(x) = \tan\left(\dfrac{\pi}{2}\cos x\right) \Rightarrow f'\left(\dfrac{\pi}{3}\right) = ?$

A) $\dfrac{-\pi\sqrt{3}}{2}$ B) $-\dfrac{\pi}{2}$ C) 0   D) π E) $\dfrac{\pi\sqrt{3}}{2}$

18. $f(x) = x \cdot \arcsin x + \sqrt{1 - x^2}$ $\Rightarrow$ $f'\left(\dfrac{1}{2}\right) = ?$

A) $\dfrac{\pi}{8}$ B) $\dfrac{\pi}{6}$ C) $\dfrac{\pi}{3}$ D) $\dfrac{\pi}{4}$ E) $\dfrac{\pi}{2}$

19. $\begin{cases} x = t^3 + 3t \\ y = t^3 - 3t \end{cases}$ $\Rightarrow$ $\dfrac{d^2y}{dx^2}\bigg|_{t=1} = ?$

A) 0   B) $\dfrac{1}{6}$ C) $\dfrac{1}{4}$ D) $\dfrac{1}{2}$ E) 1

20. $f(x) = (x^2 - x)^{-2}$ $\Rightarrow$ $f'(2) = ?$

A) -4 B) $\dfrac{-8}{3}$ C) $-\dfrac{3}{4}$ D) $\dfrac{1}{4}$ E) $\dfrac{1}{2}$

Answers					
1. A	2. A	3. B	4. C	5. D	6. E
7. A	8. C	9. C	10. B	11. D	12. D
13. E	14. A	15. E	16. C	17. A	18. B
19. B	20. C				

Test 6

1. $f(x) = \dfrac{2x^3 + 1}{x} \implies f'(x) = ?$

A) $2x - \dfrac{2}{x^3}$ B) $2 + 2x^2$ C) $2x^2 - \dfrac{1}{x^3}$

D) $4x - \dfrac{1}{x^2}$ E) $x - \dfrac{1}{x^3}$

2. $f(x) = \dfrac{x^2 + 1}{x^2 - 1} \implies f'(x) = ?$

A) $\dfrac{2x}{(x^2 - 1)^2}$ B) $\dfrac{4x}{(x^2 - 1)^2}$ C) $\dfrac{-4x}{(x^2 - 1)^2}$

$$D) \frac{-8x}{(x^2-1)^3} \qquad E) \frac{-8x}{x^2-1}$$

3. $f(x) = \dfrac{x^3}{3} - \dfrac{x^2}{2} + x - 1 \quad \Rightarrow \quad f'(2) = ?$

A) 6   B) 5   C) 4   D) 3   E) 2

4. $y = (x^3 - 3x)^4 \quad \Rightarrow \quad \dfrac{dy}{dx} = ?$

A) $4(x^2 - 3x)^4$ B) $3(x^2 - 3)(x^3 - 3x)^3$

C) $12(x^2 - 1)(x^3 - 3x)^3$ D) $12(x^2 - 3)$

E) $12(x^3 - 3x)^4$

5. $f(3x - 4) = (x^2 - 2)^3 \quad \Rightarrow \quad f'(2) = ?$

A) 3   B) 4   C) 6   D) 7   E) 8

6. $f(3x - 2) = (9x^2 + 3x - 6)^2 \quad \Rightarrow \quad \dfrac{df(x)}{dx} = ?$

A) $(2x + 5) \cdot (x^2 + 5x)$ B) $2(2x + 5)(x^2 + 5x)$

C) $4(2x + 5)^2$ D) $2x(x^2 + 5x)$

E) $4x(x^2 + 5x)$

7. $f(x) = \dfrac{(x + 1)^2}{(x^2 + 1)^3} \Rightarrow f'(0) = ?$

A) -1 B) 0 C) 1 D)$\dfrac{3}{2}$ E) 2

8. $f(x) = (x^2 + 1)^3 (x^3 - 1)^2 \Rightarrow \dfrac{df(x)}{dx} = ?$

A) $6x^2(x^2 + 1)(x^3 - 1)\cdot(2x^3 + x)$

B) $6x(x^2 + 1)^2 (x^3 - 1)^3 (2x^3 + x - 2)$

C) $36x(x^2 + 1)(x^3 - 1)\cdot(2x^3 + x - 2)$

D) $6x(x^2 + 1)^2 (x^3 - 1)\cdot(2x^3 + x - 1)$

E) $12x (x^2 + 1)^2 (x^3 - 1)\cdot(x^3 + x + 1)$

9. $x^3 + y^3 = 1 \Rightarrow \dfrac{d^2y}{dx^2} = ?$

A)$\dfrac{-x}{y^3}$ B)$\dfrac{-2x}{y^5}$ C)$\dfrac{x}{y^4}$ D)$\dfrac{4x}{y^5}$ E)$\dfrac{-4x}{y^5}$

10. $\begin{cases} x = f(t) = t + \dfrac{1}{t} \\ y = g(t) = t - \dfrac{1}{t} \end{cases}$ $\Rightarrow$ $\dfrac{d^2y}{dx^2} = ?$

A) $\dfrac{4t^3}{(t^2 - 1)^3}$ B) $\dfrac{-2t}{(t^2 - 1)}$ C) $\dfrac{-4t^2}{(t^2 - 1)^3}$

D) $\dfrac{-4t^3}{(t^2 - 1)^3}$ E) $\dfrac{8t}{(t^2 + 1)^3}$

11. $f(x) = x^2 + 1,$

$\qquad g(x) = \sqrt{x^2 + 1}$ $\Rightarrow$ $\dfrac{df(x)}{dg(x)} = ?$

A) $2f(x)$ B) $\dfrac{2}{g(x)}$ C) $g(x)$ D) $2g(x)$ E) $\dfrac{g(x)}{f(x)}$

12. $f(x) = (3x - 4) \cdot \sqrt[4]{(x + 1)^3}$

$\qquad \Rightarrow f'(x) = ?$

A) $\dfrac{x}{12\sqrt[4]{(x + 1)^3}}$ B) $\dfrac{12x - 9}{4\sqrt[4]{x + 1}}$ C) $\dfrac{21x}{2\sqrt[4]{(x + 1)^3}}$

D) $\dfrac{7x}{4\sqrt[4]{x + 1}}$ E) $\dfrac{35x}{6\sqrt[4]{x + 1}}$

13. $f(x) = \dfrac{3x^2 - 1}{\sqrt[3]{(x^3 - 1)^2}}$ $\Rightarrow$ $\dfrac{df(x)}{dx} = ?$

A)$\dfrac{2x(x-3)}{\sqrt[3]{x^3-1}}$ B)$\dfrac{x(x-3)}{(x^3-1)(\sqrt{x^3-1})}$ C)$\dfrac{2x(x-3)}{(x^3-1)\sqrt[3]{(x^3-1)^2}}$

D)$\dfrac{x-3}{(x^3-1)^2\sqrt{x^3-1}}$ E)$\dfrac{2x}{(x^3-1)\sqrt[3]{(x^3-1)}}$

14. $x^2 - xy + y^2 - 3 = 0 \implies \dfrac{dy}{dx} = ?$

A)$\dfrac{2x-y}{x-2y}$ B)$\dfrac{2x+y}{x-2y}$ C)$\dfrac{x-2y}{2y+x}$

D)$\dfrac{x+y}{x-y}$ E)$\dfrac{y-2x}{x+2y}$

15. $\arctan\dfrac{x}{y} + \ln\sqrt{x^2+y^2} = 0 \implies \dfrac{dy}{dx} = ?$

A)$\dfrac{x-y}{x+y}$ B)$\dfrac{x+y}{x-y}$ C)$\dfrac{2x+y}{2x-y}$

D)$\dfrac{y-2x}{2x+y}$ E)$\dfrac{x+y}{x-2y}$

16. $f(x) = \arcsin\dfrac{x-1}{3} \implies f'(1) = ?$

A)$\dfrac{1}{\sqrt{3}}$ B)$\dfrac{1}{3}$ C)$\dfrac{1}{3\sqrt{3}}$ D)$\dfrac{\sqrt{3}}{6}$ E)$\dfrac{1}{9}$

17. $f(x) = x \cdot \sin \dfrac{1}{x} \Rightarrow \dfrac{d^2 f(x)}{dx^2} = ?$

A) $\dfrac{1}{x^2} \cdot \cos \dfrac{1}{x}$ B) $-\dfrac{1}{x^3} \cdot \cos \dfrac{1}{x}$ C) $\dfrac{1}{x} \cdot \sin \dfrac{1}{x}$

D) $\dfrac{1}{x^3} \cdot \sin \dfrac{1}{x}$ E) $-\dfrac{1}{x^3} \cdot \sin \dfrac{1}{x}$

Answers					
1. D	2. C	3. D	4. C	5. E	6. B
7. E	8. D	9. B	10. D	11. D	12. B
13. C	14. A	15. B	16. B	17. E	

Chapter 16   **The Derivative**

Test 7

1. $f(x) = \dfrac{x^3 + 1}{x^3 + 3} \Rightarrow f'(3) = ?$

A) $\dfrac{1}{25}$ B) $\dfrac{2}{25}$ C) $\dfrac{3}{50}$ D) $\dfrac{3}{20}$ E) $\dfrac{4}{75}$

2. $f(x) = \dfrac{1 - \cos 2x}{1 + \cos 2x} \Rightarrow f'\left(\dfrac{\pi}{3}\right) = ?$

A) 3   B) $4\sqrt{3}$ C)$\dfrac{2\sqrt{3}}{3}$ D)$\dfrac{\sqrt{3}}{3}$ E) 16

3. $f(x) = \dfrac{1}{(\cos 2x - \sin 2x)^2} \implies f'\left(\dfrac{\pi}{24}\right) = ?$

A) $4\sqrt{2}$ B) $4\sqrt{3}$ C) 12   D) 16   E) $8\sqrt{3}$

4. $f(x) = \left[\dfrac{2\sin^2 x - \tan 3x}{(1 + \cos 2x)\sqrt{2 + \sec^2 x}}\right]^{-6} \implies f'\left(\dfrac{\pi}{4}\right) = ?$

A) 2   B)$\dfrac{5}{2}$ C) 3   D)$\dfrac{7}{2}$ E) 4

5. $x = \dfrac{\pi}{3}$

$f(x) = \sec x \implies \dfrac{d^2 f(x)}{dx^2} = ?$

A) 8   B) 10   C) 12   D) 14   E) 16

6. $f(x) = \arctan\left(\dfrac{4\sin x}{3 + 5\cos x}\right) \implies f'\left(\dfrac{\pi}{3}\right) = ?$

A)$\dfrac{4}{5}$ B)$\dfrac{6}{11}$ C)$\dfrac{8}{13}$ D)$\dfrac{8}{25}$ E)$\dfrac{16}{25}$

7. $f(x) = \arctan\left(\dfrac{1}{x}\right) \Rightarrow f'(2) = ?$

A)$\dfrac{1}{2}$ B)$\dfrac{1}{3}$ C)$\dfrac{1}{5}$ D)$-\dfrac{1}{5}$ E)$-\dfrac{1}{6}$

8. $f(x) = e^x \cdot \ln x \Rightarrow f'(x) = ?$

A) $e^x\left(\ln x + \dfrac{1}{x}\right)$ B) $e^x\left(1 + \dfrac{1}{x}\right)$ C)$\ln x(e^x + 1)$

D)$\dfrac{e^x}{x}$ E) $e^x + \ln x$

9. $e^{x+y} = y^x \Rightarrow \dfrac{dy}{dx} = ?$

A)$\dfrac{y}{x(x+y)}$ B)$\dfrac{y^2}{x(y-x)}$ C)$\dfrac{y^3}{x(y-x)}$

D)$\dfrac{y^3}{x^2 - xy}$ E)$\dfrac{y^2}{x^2 + y}$

10. $f(x) = \dfrac{x^3 + x + 2}{x^2 + 4}$

$\Rightarrow \lim\limits_{h \to 0} \dfrac{f(2+h) - f(2)}{h} = ?\boxed{10}$

A) 2   B) 1   C)$\frac{5}{6}$ D)$\frac{13}{16}$ E)$\frac{7}{8}$

11. $f(x) = \dfrac{x^3 + 8}{x^2 + 2}$

$\Rightarrow \lim\limits_{x \to 2} \dfrac{f(x) - f(2)}{x - 2} = ?$

A)$\frac{2}{9}$ B)$\frac{3}{5}$ C)$\frac{5}{6}$ D) 1   E)$\frac{3}{2}$

12. $f(x) = x^3 - 4|x| + 2x \Rightarrow f'(2) = ?$

A) 8   B) 10   C) 12   D) 15   E) 24

13. $f(x) = x^3 - 6x + g(4x - 7)$

$g'(5) = -3 \Rightarrow f'(3) = ?$

A) – 6   B) – 3   C) 0   D) 6   E) 9

14. $f(x) = x^2 + 1$

$g(x) = x^3 + 2x \Rightarrow (fog)'(1) = ?$

A) 30   B) 28   C) 24   D) 18   E) 16

15. $f(x) = x^3 - x^2 - 12x + 7$

$\Rightarrow (f^{-1})'(7) = ?$

A)$\dfrac{3}{16}$ B)$\dfrac{1}{86}$ C)$\dfrac{1}{28}$ D)$\dfrac{1}{24}$ E)$\dfrac{5}{96}$

16. $\begin{cases} x = 6t - 4 \\ y = t^3 + 8 \end{cases}$ $\Rightarrow f'(8) = ?$

A) – 2   B) – 1   C) 0   D) 1   E) 2

17. $f(x) = x^5 - 2x^4 + x^3 - 2x^2 + 4x - 4$

$\Rightarrow (f^{-1})'(4) = ?$

A)$\dfrac{1}{6}$ B)$\dfrac{1}{8}$ C)$\dfrac{1}{12}$ D)$\dfrac{1}{16}$ E)$\dfrac{1}{24}$

18. $f(x) = x \cdot \sqrt{x^3 + 8}$ $\Rightarrow f'(2) = ?$

A)$\dfrac{4}{9}$ B)$\dfrac{5}{2}$ C) 5   D)$\dfrac{13}{2}$ E) 7

19. $f(x) = \arctan(\sin x)$

87

$$\cos a = \frac{2}{5} \Rightarrow f'(a) = ?$$

A)$\frac{7}{24}$ B)$\frac{5}{23}$ C)$\frac{3}{20}$ D)$\frac{2}{9}$ E)$\frac{1}{25}$

Answers					
1. C	2. A	3. E	4. C	5. D	6. C
7. D	8. A	9. B	10. E	11. A	12. B
13. E	14. A	15. C	16. E	17. E	18. E
19. B					

Chapter 16   The Derivative

Test 8

1. $f(x) = \arcsin(\tan x)$

    $$\tan \theta = \frac{1}{3} \Rightarrow f'(\theta) = ?$$

    A)$\frac{5\sqrt{2}}{6}$ B)$\frac{5\sqrt{3}}{3}$ c)$\frac{5}{6}$ D)$\frac{3}{16}$ E)$\frac{4}{9}$

2. $f(x) = e^{\sin(\ln x)} \Rightarrow f'(1) = ?$

A) 1   B) 2   C)$\dfrac{1}{e}$ D) e   E) π

3. $f(x) = x^{\sin x} \Rightarrow f'\left(\dfrac{\pi}{2}\right) = ?$

A)$\dfrac{-\pi}{4}$ B)$\dfrac{-\pi}{3}$ C)$\dfrac{-\pi}{2}$ D) 1   E)$\dfrac{3\pi}{2}$

4. $y = e^x \cdot e^{\ln x} \Rightarrow \dfrac{dy}{dx} = ?$

A) $e^x(x+1)$ B)$\dfrac{e^x}{x}$ C)$\dfrac{e^x \ln x}{x}$

D) $e^x(x+1)$ E) $e^x\left(1+\dfrac{1}{x}\right)$

5. $f(x) = (e^x)^{e^x} \Rightarrow f'(\ln 2) = ?$

A) 24   B) 12   C) $4 + \ln 64$ D) $8 + \ln 256$ E) $64\ln 2$

6. $y^x = e^{x+y} \Rightarrow \dfrac{dy}{dx} = ?$

A)$\dfrac{y}{y-x}$ B)$\dfrac{y^2}{x(y-x)}$ C)$\dfrac{y^2}{y(y-x)}$

89

$$D)\frac{y^2}{y(x-y)} \qquad\qquad E)\frac{2y}{xy(1-x)}$$

7. $xy = (x+y)^2 \quad \Rightarrow \quad \dfrac{dy}{dx} = ?$

$A) -\dfrac{2x+y}{x+2y} \qquad B)\dfrac{x-y}{x+y} \qquad C)\dfrac{y-2x}{2y-x}$

$D)\dfrac{2xy}{x+y} \qquad\qquad E)\dfrac{x+y}{y-2x}$

8. $f(x) = \dfrac{2}{x^2+1} \quad \Rightarrow \quad \lim\limits_{h\to 0}\dfrac{f(-4+h)-f(-4)}{h} = ?$

$A)\dfrac{4}{17} \qquad B)\dfrac{6}{54} \qquad C)\dfrac{12}{108} \qquad D)\dfrac{16}{225} \qquad E)\dfrac{16}{289}$

9. $f(x) = \dfrac{1}{x-1}\ (x \neq 1) \quad \Rightarrow \quad \lim\limits_{t\to 0}\dfrac{f(t-2)-f(-2)}{t} = ?$

$A) -\dfrac{1}{3} \qquad B) -\dfrac{1}{6} \qquad C) -\dfrac{1}{9} \qquad D)\dfrac{1}{3} \qquad E)\dfrac{1}{9}$

10. $\begin{cases} x = t - t^3 \\ y = 1 - t \end{cases} \quad \Rightarrow \quad \dfrac{dy}{dx}\bigg|_{t=1} = ?$

A) $\dfrac{-2}{5}$ B) $-\dfrac{1}{2}$ C) 0   D) $\dfrac{1}{2}$ E) 1

11. $\begin{cases} x = t^3 - 4t \\ y = t^3 \end{cases}$ $\Rightarrow$ $\dfrac{d^2y}{dx^2}\Big|_{t=1} = ?$

A) 6   B) 9   C) 12   D) 18   E) 24

12. $\begin{cases} x = \cos t \\ y = \sin t \end{cases}$ $\Rightarrow \dfrac{d^2y}{dx^2} = ?$

A) $\tan t\sec(t)$ B) $\cot(t)$ C) $-\sec^3 t$

D) $\tan^2 t$ E) $4 - cosec^3 t$

13. $y = \log(2x + 1)$ $\Rightarrow \dfrac{dy}{dx} = ?$

A) $\dfrac{2}{2x+1}$ B) $\dfrac{2}{2x+1}\cdot \ln 10$ C) $\dfrac{2\log e}{2x+1}$

D) $\dfrac{2}{(2x+1)^2}$ E) $\dfrac{\log e}{2x+1}$

14. $y = x^{\ln 3}$ $\Rightarrow \dfrac{dy}{dx} = ?$

A) $\ln 3 \cdot x^{\ln \frac{3}{e}}$ B) $\dfrac{\ln 3}{x}$ C) $\dfrac{\ln^2 3}{x}$

D) $\dfrac{\ln \frac{3}{e}}{x}$ E) $\dfrac{1}{x^{\ln 3}}$

15. $y = \ln x^3 + \ln^3 x \Rightarrow \dfrac{dy}{dx} = ?$

A) $\dfrac{3(1 + \ln x)}{x}$ B) $\dfrac{3(1 + \ln^2 x)}{x}$ C) $\dfrac{1 + \ln^2 x}{x^2}$

D) $\dfrac{3\ln x}{x}$ E) $\dfrac{3(x + \ln x)}{x}$

16. $y = \ln [(x^2 + 2)^2 \cdot (x^3 + x - 1)] \Rightarrow \dfrac{dy}{dx}\Big|_{x=2} = ?$

A) $\dfrac{37}{9}$ B) $\dfrac{8}{3}$ C) $\dfrac{25}{9}$ D) 3   E) 4

17. $y = \ln^2 (2x + 3) \Rightarrow \dfrac{dy}{dx}\Big|_{x=3} = ?$

$$A)\frac{\ln 3}{9} \qquad B)\frac{4\ln^3}{9} \qquad C)\frac{8}{9} \qquad D)\frac{16}{27} \qquad E)\frac{8\ln 3}{9}$$

18. $y = \log_3 (3x + 1) \Rightarrow \dfrac{dy}{dx} = ?$

$$A)\frac{3\log_3 e}{3x + 1} \qquad B)\frac{3}{3x + 1} \qquad C)\frac{3\ln 3}{3x + 1}$$

$$D)\frac{\log_3 9}{3x + 1} \qquad E)\frac{27}{3x + 1}$$

19. $f(x) = (x^2 + 1)\cdot\ln (2x + 1) \Rightarrow f'(1) = ?$

$$A)\frac{\ln 81 + 4}{3} \qquad B)\frac{\ln 729 + 4}{9} \qquad C)\frac{\ln 243 + 4}{3}$$

$$D)\frac{2\,\ln⬚(27\cdot e^2)}{3} \qquad E)\frac{\ln⬚(27\cdot e^2)}{9}$$

Answers

1. A	2. A	3. D	4. A	5. D	6. B
7. A	8. E	9. C	10. E	11. E	12. E
13. C	14. A	15. B	16. C	17. E	18. A
19. D					

Test 9

1. $y = \sqrt{4 + \ln x} \quad \Rightarrow \dfrac{dy}{dx} = ?$

A) $\dfrac{1}{\sqrt{4 + \ln x}}$ B) $\dfrac{1}{x\sqrt{4 + \ln x}}$ C) $\dfrac{1}{2x\sqrt{4 + \ln x}}$

D) $\dfrac{2x}{\sqrt{4 + \ln x}}$ E) $\dfrac{x}{2x\sqrt{4 + \ln x}}$

2. $x^3 + 4xy^2 - y^4 - 27 = 0 \quad \Rightarrow \dfrac{dy}{dx} = ?$

A) $\dfrac{3x^2 - 4y^2}{4y^3 + 8xy}$ B) $\dfrac{3x^2 + 4y^2}{4y^3 - 8xy}$ C) $\dfrac{x^2 + 2y^2}{2y^2 - 4xy}$

D) $\dfrac{3x^2 + 4y^2}{4y^3 + 8xy^2}$ E) $\dfrac{3x^2 + 4y^2}{4y^4 - 8x^2y}$

3. $y = x^x \quad \Rightarrow \dfrac{dy}{dx} = ?$

A) $x^x \cdot \ln x$ B) $(\ln x + 2) \cdot x^x$ C) $x^x \cdot \ln(ex)$

D) $x^x \cdot (1 - \ln x)$ E) $x^x \cdot \ln(e^2 x)$

4. $y = e^x \cdot x^{3x} \quad \Rightarrow \dfrac{dy}{dx} = ?$

94

A) $e^x \cdot x^{3x} \cdot (4 + 3\ln x)$ B) $e^x \cdot x^{3x} \cdot (3 + \ln x)$

C) $x^{3x} \cdot (4 + \ln x)$ D) $e^x \cdot x^{3x} \cdot (1 + 4\ln x)$

E) $4 \cdot x^{3x} \cdot e^x \cdot \ln x$

5. $f(x) = x^{2x} \implies f'(2) = ?$

A) 64   B) 32   C) $16(\ln 2 + 1)$

D) $32(\ln 2 + 1)$ E) $64(\ln 2 + 1)$

6. $y^2 = e^{x+y} \implies \dfrac{d^2y}{dx^2} = ?$

A) $\dfrac{2y}{(2-y)^2}$ B) $\dfrac{y}{(2+y)^2}$ C) $\dfrac{y^2}{(2-y)^3}$

D) $\dfrac{4y}{(2-y)^4}$ E) $\dfrac{y^2 + 2y}{(2+y)^3}$

7. $f(x) = \dfrac{\sqrt{1 + 2x} - (1 + x)}{x^2} \implies \lim_{x \to 0} f(x) = ?$

A) 1   B) $\dfrac{2}{3}$ C) $\dfrac{1}{2}$ D) $-\dfrac{1}{2}$ E) $-\dfrac{1}{4}$

8. $f(t) = \dfrac{\sqrt[3]{1+3t} - (1+t)}{t^2}$

$\Rightarrow \lim\limits_{t \to 0} \dfrac{\sqrt[3]{1+3t} - (1+t)}{t^2} = ?$

A) $-\dfrac{5}{2}$ B) $-\dfrac{3}{2}$ C) $-\dfrac{1}{2}$ D) -1 E) $\dfrac{5}{2}$

9. $f(x) = \dfrac{2\sin^2 x}{1 - \cos x}$

$\Rightarrow \lim\limits_{h \to 0} \dfrac{f(h) - f(0)}{h} = ?$

A) 0 B) 1 C) 2 D) 3 E) 4

10. $\lim\limits_{x \to \frac{\pi}{2}} \dfrac{1 - \sin x}{\cos x} = ?$

A) $-\dfrac{1}{2}$ B) -1 C) 0 D) $\dfrac{1}{2}$ E) 1

11. $\lim\limits_{x \to \pi} \dfrac{\cos \dfrac{x}{2}}{x - \pi} = ?$

A) 2   B)$\dfrac{3}{2}$ C) 1   D) – 1   E) $-\dfrac{1}{2}$

12. $\lim\limits_{x \to \pi} \dfrac{1 + \cos x}{\sin^2 x} = ?$

A) $-\dfrac{1}{2}$ B) 0   C)$\dfrac{1}{2}$ D)$\dfrac{1}{4}$ E)$\dfrac{1}{8}$

13. $\lim\limits_{x \to 0} \sin 4x \cdot \cot 2x = ?$

A)$\dfrac{1}{24}$ B)$\dfrac{1}{12}$ C)$\dfrac{1}{4}$ D) 2   E) 3

14. $\lim\limits_{x \to 0} \dfrac{4 \sin 9x}{3x} = ?$

A)$\dfrac{1}{12}$ B)$\dfrac{1}{6}$ C) 6   D) 9   E) 12

15. $\begin{cases} y = t^3 + t \\ x = t^3 + 1 \end{cases} \Rightarrow \dfrac{d^2 y}{dx^2}\Big|_{x = 2} = ?$

A) $-\dfrac{2}{9}$ B) $-\dfrac{1}{18}$ C)$\dfrac{2}{3}$ D)$\dfrac{1}{6}$ E)$\dfrac{2}{9}$

16. $\lim\limits_{x\to\frac{\pi}{4}} \dfrac{\sin x - \cos x}{\cos^2 x - \sin^2 x} = ?$

A) $-\sqrt{2}$ B) $-\dfrac{\sqrt{2}}{2}$ C) $-\dfrac{1}{2}$ D) $\dfrac{\sqrt{2}}{2}$ E) $\sqrt{2}$

17. $\lim\limits_{x\to 2} \dfrac{\ln x - \ln 2}{x^2 - 4} = ?$

A) 1 B) $\dfrac{1}{2}$ C) $\dfrac{1}{4}$ D) $\dfrac{1}{8}$ E) $\dfrac{1}{16}$

18. $e^{-x^3}\dfrac{d^2}{dx^2}\left(x^2 e^{x^3}\right) = ?$

A) $9x^6 + 18x^3 + 2$ B) $9x^6 + 6x^4 + 12x^3 + 2$

C) $12x^5 + 9x^3 + 4$ D) $18x^6 + 9x^3 + 2$

E) $2x^6 + 4x^5 + 12x^4 + 9x^3$

19. $f(x) = \sin 6x \cdot \cos 6x \Rightarrow f'\left(\dfrac{\pi}{12}\right) = ?$

A) -12 B) -6 C) 0 D) 6 E) 12

20. $\lim\limits_{x \to \pi} (x - \pi) \cdot cosec\ x = $?

A) -1 B) $-\dfrac{1}{2}$ C) 0 D) $\dfrac{1}{2}$ E) 1

21. $y = \sin\left(\dfrac{2x - 1}{x + 1}\right)$

$x = \dfrac{\pi + 2}{4 - \pi} \quad \Rightarrow \quad \dfrac{dy}{dx} = $?

A) $\dfrac{(4 - \pi)^2}{12}$ B) $\dfrac{(4 + \pi)^2}{18}$ C) $\dfrac{(2 - \pi)^3}{6}$

D) $\dfrac{\pi + 2}{6}$ E) $\dfrac{\pi - 4}{6}$

Answers					
1. C	2. B	3. C	4. A	5. D	6. A
7. D	8. C	9. E	10. C	11. D	12. A
13. D	14. E	15. A	16. B	17. D	18. A
19. B	20. A	21. A			

www.ingramcontent.com/pod-product-compliance
Lightning Source LLC
Chambersburg PA
CBHW072156170526
45158CB00004BA/1676